to Timothy

Our Queen's Jubilee 1977

from Kay Drury

A Salute to BRITISH GENIUS

A Salute to
BRITISH GENIUS

by
Gordon
Rattray Taylor

Secker & Warburg/London

First published in England 1977 by
Martin Secker & Warburg Limited
14 Carlisle Street, London W1V 6NN

Copyright © John Player Foundation 1977

SBN 436 51639 X

Designed by Philip Mann

Printed in Great Britain by
Westerham Press Limited
Westerham, Kent

Illustration acknowledgments

Grateful acknowledgment is due to the following for their permission to reproduce illustrations: Agricultural Research Council, 92; Andrews McClaren Ltd, 149; Charles Blatchford & Sons Ltd, 135; Bristol University, Department of Anatomy, 146; British Aircraft Corporation Ltd, 97, 117; British Airways, 81, 90; British Hovercraft Corporation Ltd, 59; British Library, 154; British Musem Newspaper Library, 34; British Petroleum, 86; British Rail, 81, 90; British Steel Corporation Ltd, 91; Cambridge University, Autarkic Housing Project, 140; Cavendish Laboratory, Cambridge, 61, 152; Central Electricity Generating Board, 101; Central Press Photos Ltd, 51; I.F. Clarke Collection, 16–17; Courtaulds Ltd, 33; Dunlop Ltd, 120, 150, 153; Edinburgh University, Department of Chemical Engineering, 130; Electric & Musical Industries Ltd, 93; Ferranti Ltd, 104–5; Fisons Ltd, 106; *Flight International*, 59; Glasgow Museums & Art Galleries, 76; Goblin Ltd, 81; Hawker Siddeley Aviation Ltd, 90; Hawker Siddeley Dynamics Ltd, 156; *Illustrated London News*, 14; Imperial War Museum, 44, 45, 46, 47, 51, 52, 53, 54, 55, 56–7; Independent Television News Ltd, 147; International Research & Development Co Ltd, 151; The Mathilda & Terence Kennedy Institute of Rheumatology, 136; Keystone Press Agency Ltd, 65; Lanchester Polytechnic, 41, 42; Leyland Cars, 64; M.C.C., 22; Malaysian Rubber Co, 81; Mansell Collection, 24–5; Marconi Co Ltd, 49; Minton Museum, Royal Doulton Tableware Ltd, 20; Montagu Motor Museum, 48; Museum of London, 15, 20; National Coal Board, 100; National Engineering Laboratory, Kilbride, 124; National Film Archive, 67; National Gallery of Canada, Ottawa, 73; National Institute of Agricultural Engineering, 107, 114, 116; National Physical Laboratory, 85; National Portrait Gallery, London, 31, 36, 68, 69, 71; National Railway Museum, York, 48; Netlon Ltd, 124; *New Scientist*, 98; Patscentre International, 148; Pilkington Ltd, 125; Plant Breeding Institute, Cambridge, 108; The Plessey Company, 94; The Post Office, 118; Robert L. Priestley Ltd, 152; Radio Times Hulton Picture Library, 43, 52, 58, 65, 70, 77, 78, 79, 80; Rolls-Royce (1971) Ltd, 121; R.A.C. Tank Museum, Bovington Camp, 45; Royal College of Physicians, Edinburgh, 136; Royal Geographical Society, 58; Royal Signals and Radar Establishment, Malvern, 54, 123; The Royal Television Society, 50; St Bride Printing Library, 74–5; St Mary's Hospital Medical School, 63; Science Museum, London: Crown copyright, 23, 30, 40, 84, photo, 26, 27, 28, 30, lent by the late Sir J.J. Thomson, 37, lent by the late Sir Oliver Lodge, 38, 39; Smallford Planters Ltd, 112; Spode Museum, Spode Ltd, 21; Standard Telephones & Cables Ltd, 87, 119, 145; Stanmore, Institute of Orthopaedics (University of London), Royal National Orthopaedic Hospital: Department of Biomedical Engineering, 133, 134, Experimental Pathology Unit, 136; The Tate Gallery, London, 72; Transhore International Ltd, 147; Unilever Ltd, 95, 110; U.K.A.E.A., 83, 96; Wates Ltd, 142; Wedgwood Ltd, 21; Westland Aircraft, 90; Wind Energy Supply Co Ltd, 83. Thanks are due also to the following photographers: Dr S.Y. Ali, 136; Csaky/Schneebeli, 81, 82, 83, 84, 87, 88, 89, 91, 100, 103, 124, 127, 133, 139, 146, 149, 152, 155; Dr P.A. Kendall, 136; Eileen Tweedy, 13, 20, 46; Eric Wilmot, 83, 96; and to Cartographic Enterprises for artwork, 101, 102.

For generations past British people have
contributed more than a fair share to the advance
of human civilisation. This exhibition displays
the material achievements of the last hundred
years, certainly the most fruitful period in our
long history and probably the most prolific output
by any nation in a comparable period of time.

That this genius has not dried up is
demonstrated by the number of brilliant ideas of
our own generation, but if they are to flourish
they must be planted in a fertile soil.

The John Player Foundation has done an
important service to the nation by sponsoring
this exhibition. I hope it will be an encouragement
and an inspiration to all who see it.

1977.

Contents

Preface

This year we at John Player and Sons are celebrating our Centenary. When a young man named John Player, a former draper's assistant, took over a tobacco factory in Nottingham in 1877, he could not have imagined the sort of world in which the business would be operating a hundred years later – a world altered by economic, social and technological changes beyond the wildest dreams of Mr Player and his contemporaries.

The last century has been marked by unparalleled scientific discovery and innovation. The real success has been in harnessing these developments to provide tangible benefits for man. The period has produced vast improvements in living standards, which in turn have meant fuller lives for individuals and a greater degree of civilisation for many countries.

In that time, Britain's achievements have been formidable despite two world wars, major international commitments and adverse economic conditions, which have imposed a severe strain on our resources, both human and material.

These advances have of course also stimulated great social changes, many of which are now making themselves felt, while others are still to come. I am afraid that gloom is one of the more fashionable moods evoked by the problems inherent in these changes, one which I believe to be largely unwarranted – it should not be allowed to obscure past, present and prospective British achievement. That is why we at John Player have, through the John Player Foundation, sponsored the British Genius Exhibition in 1977.

Our conscious, if ambitious, objective has been to excite the imagination of the whole British public. If this Exhibition and this book help in any way to restore confidence in our future then we at John Player will be well satisfied that our Centenary has been appropriately celebrated.

Geoffrey Kent

Foreword

This book of the British Genius Exhibition is meant to remind us that we are an extremely resourceful and inventive people. As you read it, we hope it will make you feel – as it makes me feel – that what we have done before, we can perfectly well do again. Not only did we start the Industrial Revolution two hundred years ago, pulling ourselves up by our bootstraps and the world after us, but we have maintained that impetus right up to the present. The inventions of the last twenty-five years have been as prolific as at any time in our history. And looking to the future I am sure that the strain of British genius is as strong as ever.

This inventive genius has, however, been overshadowed by economic problems as we realised our transition, surely but painfully, from Empire to Europe; the setbacks and the internal strains they caused have made us less self-confident. Now we need to look ahead again. As the old prophet said, "Where there is no vision, the people perish". We need to rediscover national objectives and regain our confidence, our belief in ourselves and in our ability to rediscover our national genius. This is what the John Player Foundation British Genius Exhibition and this book are all about.

Gordon Rattray Taylor, author of this "Salute to British Genius", argues that we have possibly contributed more to the advance of the world than any nation since the Greeks, and that in the modern world our contribution is greater in absolute terms. So we should not be too downhearted.

Of course we still have problems, but we also retain our genius for finding a way round them. We have a major problem of resources, of finding funds to make the investment which alone can translate the inventive genius of the British into marketable products. I am sure that we will find a way round this problem as the oil revenues roll in. We will have a problem of replacing the energy when the oil runs out – and we are already full of bright ideas for doing that. We have the social problems of a multi-racial society, which our genius for tolerance will have to solve. We have problems of violence and alienation with loss of religious faith, but John Wesley faced similar problems in the eighteenth century, and the novels of Charles Dickens tell us of those faced by reformers such as Lord Shaftesbury in mid-nineteenth-century Britain. There are no easy answers, but a belief that problems can be solved makes their solution easier. At least we had no problems in financing the British Genius Exhibition. That was made possible by the extraordinary generosity of the John Player Foundation and we are grateful to them, and especially to their Chairman, Geoffrey Kent.

Frederick Catherwood

Acknowledgments

The British Genius Exhibition, for which this is the official souvenir book, could not have been mounted at all, let alone in eighteen months from conception to opening, without the dedicated efforts of many people. Unfortunately in the space available it is possible to mention only a few of them.

The Exhibition has been financed by the John Player Foundation, formed by John Player and Sons in their centennial year as a non-profit making body. This is the Foundation's first exhibition, and on Geoffrey Kent, Chairman of the John Player Foundation, has fallen the greatest weight. The decision to proceed and to place at risk a very large sum of money was in itself a great act of faith on his part and on that of his fellow directors. Quite apart from this, it is certain that without his calmness, cheerfulness and enthusiasm the project could not have been brought to fruition so swiftly.

Mr Kent was supported on the Exhibition Policy Committee by Dr David Edmond, who also sat on the Executive Committee. On the latter the John Player Foundation was also represented by Michael Brayley, who bore the brunt of the day-to-day administrative decisions. Many other members of John Player and Sons have helped. I have space to mention only John Andrews, Budget Accountant, who set up and operated financial systems, and Frank Doubleday, the Design Manager, who with his studio worked on the Exhibition graphics.

In the later stages of what was a very complicated project from a design and production point of view, we were helped enormously by Imperial Group Technical Services, notably by J. A. S. Burn, F. S. Townsend and G. H. Nixon.

At a very early stage Mr Kent asked Sir Frederick Catherwood to be Chairman of the Exhibition and of the Policy and Executive Committees. He has throughout been a source of inspiration and strength, and has helped to solve many apparently insurmountable problems.

A very special word of thanks is due to the Greater London Council, and in particular to the Chairman, The Lord Ponsonby of Shulbrede, who was also a most active and helpful member of the Policy Committee. James Kennedy, Head of the GLC's Parks Department, played an important part in bringing the project to fruition, as did Leslie Franklin, Features Manager, who sat on the Executive Committee. Officers of the London Borough of Wandsworth also gave much support, as did B. A. Cox, District Surveyor, Battersea District.

The response to requests for loans to the Exhibition from industry, research establishments and museums has been generous. Loans are acknowledged wherever possible within the Exhibition itself, and in the brief guide to it.

Gordon Rattray Taylor, the author of this book, was also Editorial Adviser to the Exhibition and was responsible for the development of much theme material. In November 1976 Dr Tom Margerison was appointed Executive Editor, and he took the original concept and developed it into its final form. Dr Margerison worked under enormous pressure with great initiative, intelligence and understanding. He inspired an editorial/research team whose members were Posy Harvey and Ian Marter, with Pat Carson, Tim Eiloart and Marianne Ford. Patricia Kershaw was the Research Co-ordinator.

The John Player Foundation invited Pentagram to design the Exhibition. The designers involved were Theo Crosby and Alan Fletcher, with Marvin Shane and Julia Alldridge, assisted by Peter Cumming and Kathy Tilney. This complex exhibition, involving the construction of a large temporary building, provided Mr Crosby and his colleagues with many problems. They have surmounted them with much flair.

One of the main features of the Exhibition is the multi-screen presentation devised and produced by Mick Csaky of Photophonics (a member of the Viscom Group). Mr Csaky has in addition worked tirelessly on other visual aspects of the Exhibition and has given much valuable general guidance.

Everetts are advertising agents to the Exhibition and their Managing Director, Charles Plouviez, was a most active member of the Executive Committee.

The souvenir book has been published on behalf of the John Player Foundation by Secker & Warburg, whose Managing Director, T. G. Rosenthal, and Managing Editor, Sophia Macindoe, we wish to thank for their interest and help at all times. Secker & Warburg produced the book for Carlton Cleeve, ably assisted by Candida Hunt who was also responsible for other special publications on sale at the Exhibition. Merchandise was selected by Elizabeth Malarkey.

Finally I should like to thank my co-Director David Wynne-Morgan, Chairman of Partnerplan. He was inspired to take the theme of Peter Grosvenor and James McMillan's book *The British Genius* and suggest to John Player and Sons that it might form the basis of an exhibition to mark their centenary. We worked together to turn the idea into reality. As Directors we took joint responsibility to the John Player Foundation for all aspects of the Exhibition. In practice Mr Wynne-Morgan has worked primarily on the marketing and publicity aspects while I concentrated on organisation. Mr Wynne-Morgan was assisted by Eileen Graham, who acted as Press Officer.

Philip Taverner

Architects and Designers
Theo Crosby and Alan Fletcher, Pentagram Design

Landscaping
Consultant Moyra Burnett
Contractors Doe Contracts Limited

Building
General Contractor Higgs and Hill Building Limited
Engineers Buro Happold Consulting Engineers
Partner in Charge Ian Liddell
Quantity Surveyors M. K. Boyden & Company
Services Consultants Max Fordham & Partners
Tent Clyde Canvas Goods and Structures Limited

Interior
Contractor Russell Brothers (Paddington) Limited
Models Keith Reeves, Ralph Selby, Derek Freeborn
Thorp Modelmakers Limited
Consultant to Autarkic House Alexander Pike
Ceiling to Auditorium Cape Building Products

Audio-visual Show
Producer/Director Mick Csaky
Photographs Heine Schneebeli
Research Liz Clasen
Editor Darryl Johnson
Script Consultant Betty Isaacs
Rostrum Photography Vincent Joyce
Production Assistant Rosemary Gorman
Control System Electrosonic Limited
Musical Score Tim Souster

Exhibition Management
Pauline Kennedy
Mansel Bebb

A List of British Nobelmen—and Women

Since Alfred Nobel founded his awards for work on Physics (P), Chemistry (C), and Physiology or Medicine (P & M) in 1901, some 56 British or British-naturalised scientists have won or shared an award. (N.B. A fuller list, giving details of the work done, is available in *The British Genius* by Peter Grosvenor and James McMillan.)

1902 Sir Ronald Ross (P&M)
1904 Lord Rayleigh (P) and Sir William Ramsay (C)
1906 Sir Joseph Thomson (P)
1908 Lord Rutherford (C)
1915 Sir Henry Bragg and Sir Lawrence Bragg (P)
1917 Charles G. Barkla (P)
1921 Frederick Soddy (C)
1922 Francis Aston (C) and A. V. Hill (P&M) shared with Otto Meyerhof
1927 C. T. R. Wilson (P) shared with A. H. Compton
1928 Sir Owen Richardson (P)
1929 Sir Arthur Harden (C) shared with H. von Euler-Chelpin; and Sir Frederick Gowland Hopkins (P&M) shared with C. Eijkman
1932 Sir Charles Sherrington and Lord Adrian (P&M)
1933 P. A. M. Dirac (P) shared with E. Schrödinger
1935 Sir James Chadwick (P)
1936 Sir Henry Dale (P&M) shared with O. Loewi
1937 Sir George Thomson (P) shared with C. J. Davisson; and Sir Walter Haworth (C) shared with Paul Karrer and R. Kuhn
1945 **Sir Alexander Fleming, Lord Florey and Sir Ernst Chain (P&M)**
1947 Sir Edward Appleton (P) and Sir Robert Robinson (C)
1948 Lord Blackett (P)
1950 C. F. Powell (P)
1951 Sir John Cockcroft and E. T. S. Walton (Irish) (P)
1952 A. J. P. Martin and R. L. M. Synge (C)

1953 Sir Hans Krebs (P&M) shared with Fritz Lipmann
1954 Max Born (P) shared with Walter Bothe
1956 Sir Cyril Hinshelwood (C) shared with N. Semenov
1957 Lord Todd (C)
1958 Frederick Sanger (C)
1960 Sir Peter Medawar and Sir Frank Macfarlane Burnet (Australian) (P&M)
1962 Francis Crick and Maurice Wilkins (P&M) shared with James Watson; and Max Perutz and Sir John Kendrew (C)
1963 Sir Alan Hodgkin, A. F. Huxley and Sir John Eccles (Australian) (P&M)
1964 Dorothy Hodgkin (C)
1967 Ronald Norrish and Sir George Porter (C) shared with Manfred Eigen
1969 D. H. R. Barton (C) shared with Professor Odd Hassel
1970 Sir Bernard Katz (P&M) shared with U. van Euler and J. Axelrod
1971 Dennis Gabor (P)
1972 Rodney Porter (P&M) shared with G. Edelman
1973 Nikolaas Tinbergen (P&M) shared with K. Lorenz and K. von Frisch; and Geoffrey Wilkinson (C); and Brian Josephson (P)
1974 Sir Martin Ryle and Antony Hewish (P)
1975 J. W. Cornforth (C) shared with V. Prelog

One
A Peculiar People

In spite of their hats being very ugly, Goddam! I like the English.
J. P. Béranger, a Frenchman

We British are, it is well known, a modest people. We prefer to understate, to dismiss our achievements with a deprecating "not bad". But though we may not care to say so, privately we are convinced that what is British is best. A hunt through the pages of the daily press discloses our conviction that British women are the most beautiful in the world; the British judicial system is the fairest; the Civil Service is the least corrupt; the British system of education is the best; the level of public hygiene is superior to that of all other countries; and even that our postal service is the most efficient in the world. The Dutch historian, G. J. Renier, noted in the 1930s: "Apart from a relatively small minority, the English are convinced that they are, that they own, and that they produce all that is best in the world. As the late Earl of Birkenhead has put it, they feel, almost to a man, that they lead the world in matters of moment."

Now, forty years later, this sublime confidence is shakier. A mood of apathy, even sometimes of pessimism, prevails in many quarters. But countries, like people, go through periods of lowered vitality, when everything seems too much. At such moments there is therapeutic value in a little praise, a little flattery even, and a rational assessment of one's underlying strength. Let us therefore put aside false modesty and for once enjoy the luxury of boasting a little.

In 1877, as Queen Victoria was declared Empress of India, Britain stood at the peak of her influence as a world power. In the century that followed she gave away an empire, without bloodshed. It was not taken from her. For the Empire was not held together by force but by an idea. With characteristic realism Britain had always worked towards self-government in her possessions, which demonstrates how little interested she was in power for power's sake. Today, when it is fashionable to depict nineteenth-century Britain as crudely imperialistic it is worth remembering this. What seemed and seems to some a series of defeats was more truly evidence of British pragmatism and her readiness to adapt to a changing world picture.

If Britain's political influence has declined, her influence in literature, science, technology and social administration has not. Rather has she been a leader in these spheres. In the following pages, I shall try to give some impression of what she has contributed to human well-being.

Taken for Granted

The British actually are a remarkable people. In proportion to our population, we have arguably contributed more to the advance of the world than

Victoria becomes an Empress: the Imperial Durbar at Delhi, from *The Illustrated London News* of 10 February 1877

any nation since the Greeks – and in the modern world our contribution is absolutely greater. This is a fact which we are in danger of forgetting. A generation is growing up which has little knowledge or appreciation of its extraordinary history. This is tragic, for a sense of history is the foundation of culture.

Foreigners, too, frequently have a distorted picture. It is not only that they think of England as a land of fogs and are amazed when the sun shines; often they think of us as a nation which has grabbed, a ruthless imperialist power, and know little of what we have given to the world. According to that Dutch historian "the British nation is the least understood and the most extensively misrepresented in the world".

We British tend to take for granted that so many of the things around us have been invented in Britain. We travel in trains (invented by a Cornishman), watch television (invented by a Scotsman) and light our pipes with matches (invented by a Yorkshireman). When we are ill we may be given penicillin (Fleming and Chain). When we are well we may play cricket or football – and what could be

Traffic jams of the 1870s, seen on
London Bridge by Gustave Doré

15

more British than that? So taken for granted is the British contribution that even when British personal names have become part of the language we scarcely notice it. We wear Macintoshes and Cardigans, cover ourselves at night with Blankets (Thomas Blanket of Bristol is quite forgotten), ride on McAdam roads and munch Sandwiches, without stopping to think of them as British inventions.

However, it is not only in the field of invention that the British genius has manifested itself. How many people know that the "flower border" of British gardens is a British notion – Gertrude Jekyll and William Robertson evolved it less than a century ago. Indeed, landscape gardening is one of Britain's great cultural contributions.

Then again, the modern novel, as well as modern drama, is predominantly an English creation. The British sense of humour (though Americans have denied that we have one) sells books and films worldwide. The composer Berlioz was struck by the amount of music "consumed" in England when he visited London in the middle of the nineteenth century. He would be even more impressed today. The columns of the quality papers reveal a list of concerts far longer than one would find in Paris, New York, Rome or Moscow. But the British contributions to the arts and sciences are so numerous that they call for sections to themselves. I shall not attempt to survey them here.

Britain has a remarkable record in civilised behaviour. Britain, though the home of fox-hunting, led the world in opposing cruelty to animals. Britain rebuilt the farms of her enemy after the Boer war. Samuel Plimsoll campaigned against the overloading of ships, as Wilberforce had campaigned against slavery before him. Britons conceived the Boy Scout movement and the Salvation Army. Trade unions were born in Britain; as was the Co-operative movement as we now know it. Britain led the world in emancipating women – despite a myth to the contrary, they were less restricted under Queen Victoria than their Continental sisters, as foreign visitors observed with surprise. (Max O'Rell reported that the Englishwoman "goes out without her mamma or her maid, gives you a hearty grasp of the hand, and looks you unblushingly in the face. Unmarried, free as air, she may go to a theatre take a walk or even a journey with male companions . . . A young girl of fifteen travels alone. I know some who come thus to school

The development of electricity: the street traffic of the future, an early prediction.

in London from the north of Scotland. In France, a young lady would not go without her maid to buy herself a pair of gloves in a shop on the opposite side of the street." And that was in the 1870s.)

These are things to remember at a time like the present when things are going badly. Despondency, frustration or a dogged grin-and-bear-it are understandable responses. But in moments of reverse and set-back it is well to take a longer view, to see the situation in its true historical proportions. That is one of the purposes of this book and of the British Genius Exhibition which it accompanies. The genius needed to solve our problems is still there, alive and well and operational.

At the same time, there are dangers in living in the past. We must look towards the future, analyse our problems and mobilise our assets so as to deal with them effectively. That is our second purpose.

For the future is what we make of it. From the point at which we stand we can move in many possible directions. So at every point there are options. Not only government policy but a million individual decisions will determine which path we take and what kind of future will become our Now. In the last analysis, a lifestyle has to be chosen, or perhaps adopted without conscious choice. The most fundamental of these choices is, I believe, the choice between a growth society – a society in which material possessions are the primary aim – and a society in which personal fulfilment takes precedence over possessions. Do we want earlier retirement and a shorter working week or higher wages and more possessions?

The fulfilment society has been called the post-industrial society. There are signs that the British people are already voting for such a society. There is growing interest in making work interesting. Unsocial hours of work are no longer seen as inevitable accompaniments of certain jobs. Considerations of commercial profit are increasingly put second to security of tenure or other demands. Health and safety are given more and more attention, both in the industrial context and in life generally.

But within the framework of this general attitude other more technical choices face us. Shall we attempt to meet the energy shortage by developing the coal industry or by taking on the hazards associated with nuclear power? Shall we seek a greater independence from imported food? Shall we pin our hopes to an energy-intensive agriculture or develop a more diversified and smaller-scale kind of farming? Choices just as radical face us in such fields as transport and health. Do we want more motorways? Do we want a hospital-based health service? In this book, various contributors attempt to present some of these issues.

But if the future presents problems it also presents tools for the solution of those problems. A major part of this tribute to British Genius comprises a survey of what British ingenuity is at work on at this moment. What ideas and inventions are in the pipeline? How will they affect our future?

All these threads are woven together in the home. Private houses consume one quarter of all the energy produced in Britain and consume most of the food. Communications come into the home; here our education starts; here we store our clothes, pursue our hobbies, and sustain our most important personal relationships. So it is fitting not only to choose a house of some kind as a nodal point in the exhibition but also to say something about the future of home life in this book.

The British are undeniably ingenious, but one crucial question requires attention: are we providing the right conditions for this inventiveness to find expression? It is almost a cliché that Britons invent things and other countries develop them to profitability. Is this true and if so, why? This question we shall also discuss.

British Genius?

Is there really such a thing as a *British* genius? The Brutish Isles (to use an old spelling) contain an amalgam of ethnically diverse groups, and foreign visitors have noted how widely they differ. A Dutchman long resident in England has contrasted the "proud, intelligent, religious and unfathomable Scots" with the "minute, musical, clever and temperamental Welsh", the "charming, untruthful, bloodthirsty and unreliable Irish", and the "unintellectual, restricted, stubborn, steady, pragmatic, silent and reliable English".

But the strains are much intermixed and it is just this intermixture which gives the British genius its unique character – if by genius we mean what is uniquely creative in the British personality. According to Peter Grosvenor and James McMillan, in their book *The British Genius*, "The British character is an amalgam of Puritan thrift, sobriety, self-help with not a little humbug, and a joyous abandon[ment] of restraints when the puritanism becomes too irksome. Not by accident did England pioneer the miniskirt while retaining stern licensing laws for the sale and consumption of liquor. There is Celtic feyness mixed with shrewd Nor-

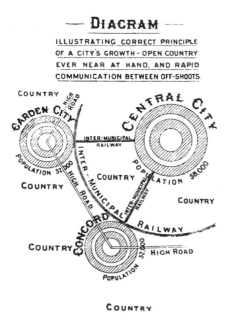

The first Garden City is born at Letchworth: Ebenezer Howard's plans of 1898

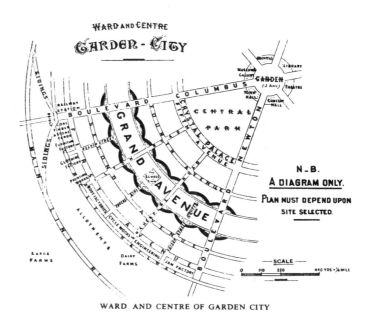

WARD AND CENTRE OF GARDEN CITY

thern calculation, Irish wit and gaiety, mingled with Scottish practicality, the English sense of duty and service which still lingers on from the Victorian Christian public school tradition . . ." Though the Puritan streak certainly has a far older provenance than Dr Arnold, maybe the general proposition is correct. British genius comes from the subtle interweaving of diverse ethnic and cultural strands.

Foreign visitors, for at least two centuries, have agreed that the English are "exquisitely polite and amazingly hospitable to strangers". They almost always said "English" though many of those they met were British rather than English. More to the point here, foreigners have always found us highly pragmatic – much more interested in whether a thing worked or not than in sticking to a theoretical principle or a logical position. In England, a Labour party which has made pacifism its programme does not hesitate to enter into coalition with a government which intends to prosecute a war. It was this pragmatism which gained for England a reputation for perfidy and hypocrisy on the Continent.

British pragmatism has certainly played a major role in British achievement, whether in the social field – British colonial administration, for instance, was highly pragmatic – or in the technological field. Thus William Thomson, Lord Kelvin, devoted many years to the designing and laying of an Atlantic telegraph cable but failed to develop the theory of electro-magnetism which was implicit in his work.

Another important strand in British character has been the high value placed on "fairness" – the word, in this precise sense, does not exist in other languages. Linked with it was a spirit of tolerance. If England is noted for her eccentrics it is no doubt because eccentricity is rather admired. The German Cohen-Portheim observed that the British like things – whether men, animals, plants or objects – to be perfect specimens of their kind, rather than that they should conform to some preconceived ideal. You find this in their gardens; it lies behind their success as stockbreeders and sportsmen; it accounts for them preferring a foreigner to look foreign and not like an imitation Englishman. It is, he said, "an attitude fundamentally different from that of the continental nations". Toleration springs from a recognition of the inevitability of individual differences, that is, an acceptance of inequality.

British belief in the importance of individual freedom and the toleration of those with unfamiliar ideas is, I fancy, the main foundation on which their achievement has been built, not only in the field of technology but in the social sphere and the arts besides. In the chapters which follow we shall see whether this is true.

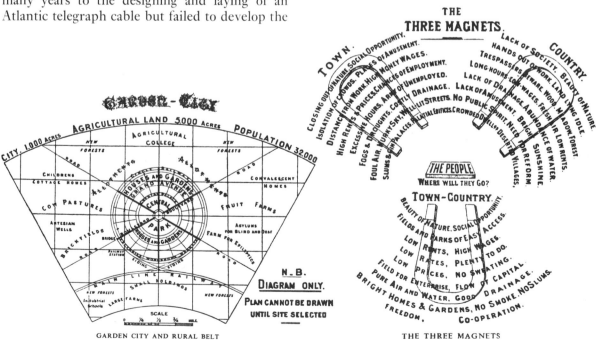

GARDEN CITY AND RURAL BELT

THE THREE MAGNETS

Two
A Century of Genius

We shall be able to construct machines which will propel ships with greater speed than a whole garrison of rowers, and which will need only one pilot to guide them. We shall be able to propel carriages with incredible speed without the assistance of an animal, and we shall be able to make machines which by means of wings will enable us to fly like birds.

The monk *Roger Bacon*, in the year A.D. 1240

Coronation mugs of 1. Victoria,
2. Edward VII, 3. Edward VIII,
4. George V, 5. George VI,
6. Elizabeth II; and, 7.,
Elizabeth II's Silver Jubilee mug

Mothers of Invention
1877–1902

The first English Test team
played Australia in March 1877

It was the best of times; it was the worst of times. Britain owned a mighty empire, on which the sun never set. Rickets was known as "the English disease" – it was due to malnutrition. British goods were exported all over the world. Women had no vote; children worked long hours. Since 1877, the year in which our retrospect of British genius starts, we have seen the world transformed.

Then the social scene looked very different from today. A visitor to London found no underground railways, no telephones, nor cars. The streets were lit by gas and paved with clatter-making stone "setts". He could not buy a postal order or even post a parcel. He would notice, too, the many barefoot and ragged children, the dirt and the fog, the stink of the Thames. Contrasting this picture with London today, you may be surprised to find how many of the changes which have taken place are due to British genius, British skill and British public spirit.

Annus Mirabilis

As it happens, the year 1877 was a particularly productive and interesting one. For one thing, it saw the first Test Match and the first tennis championships at Wimbledon. It was also the year in which Joseph Swan, with his assistant C. H. Stearn, renewed his long-abandoned efforts to produce a practical electric light bulb: he succeeded within a year, thus launching electric lighting as we know it, and thereby the whole electricity industry. It was the year in which ball-bearings were first fitted on bicycles and which saw the first steam-trams. The British Navy acquired its first destroyer. The first models of Bell's newly invented telephone reached England (though Scottish, Bell was living in Canada) and were demonstrated to the British Association at Plymouth. In science, Lord Rayleigh published his *Theory of Sound*, while Eadweard Muybridge, seeking to find new ways of using the photographic camera, produced a panoramic view of San Francisco Bay, 27 feet in length.

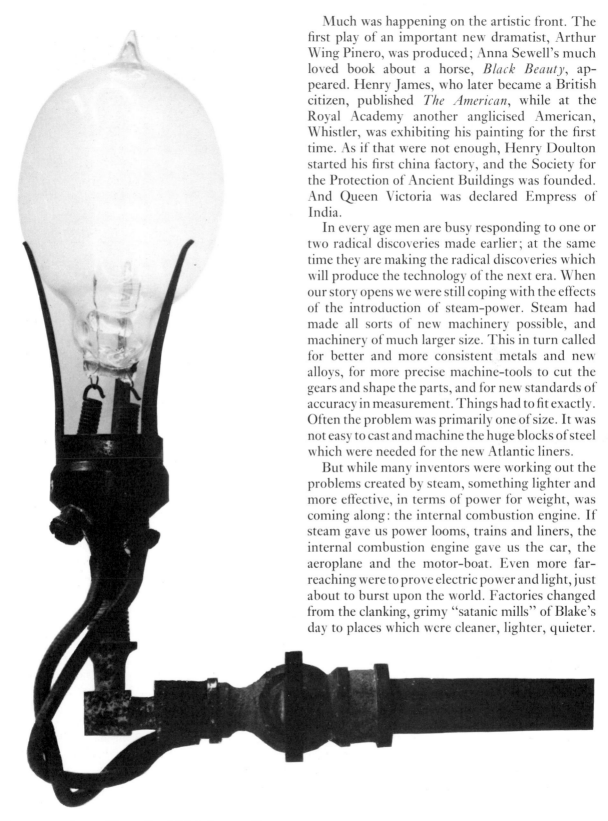

Much was happening on the artistic front. The first play of an important new dramatist, Arthur Wing Pinero, was produced; Anna Sewell's much loved book about a horse, *Black Beauty*, appeared. Henry James, who later became a British citizen, published *The American*, while at the Royal Academy another anglicised American, Whistler, was exhibiting his painting for the first time. As if that were not enough, Henry Doulton started his first china factory, and the Society for the Protection of Ancient Buildings was founded. And Queen Victoria was declared Empress of India.

In every age men are busy responding to one or two radical discoveries made earlier; at the same time they are making the radical discoveries which will produce the technology of the next era. When our story opens we were still coping with the effects of the introduction of steam-power. Steam had made all sorts of new machinery possible, and machinery of much larger size. This in turn called for better and more consistent metals and new alloys, for more precise machine-tools to cut the gears and shape the parts, and for new standards of accuracy in measurement. Things had to fit exactly. Often the problem was primarily one of size. It was not easy to cast and machine the huge blocks of steel which were needed for the new Atlantic liners.

But while many inventors were working out the problems created by steam, something lighter and more effective, in terms of power for weight, was coming along: the internal combustion engine. If steam gave us power looms, trains and liners, the internal combustion engine gave us the car, the aeroplane and the motor-boat. Even more far-reaching were to prove electric power and light, just about to burst upon the world. Factories changed from the clanking, grimy "satanic mills" of Blake's day to places which were cleaner, lighter, quieter.

There was a popular song "Gas Light, Bude Light soon will pass away/All must take a back seat to the New Electric Light":

Swan's light bulb mounted in a gas bracket, c. 1884

Streets, theatres and homes began to glow with light. And these developments sprang from the apparently useless scientific studies of men like Humphry Davy and Michael Faraday half a century before. (Faraday, when asked by royalty what was the use of his discoveries, replied: Of what use is a baby?) And likewise, while these technological advances were taking place, scientists were pottering away in laboratories, passing electric currents through gases at reduced pressures and exploring the curious effects produced by an electric spark. From these "useless" activities sprang a third great wave of technological advance: the improvement of communication, giving us radio and television, and eventually the tremendous two-edged tool of nuclear energy.

Forth Bridge – 38,000 tons of steel – as it appeared on 9 July 1889 during its construction: cantilevers had never been used on a major bridge before

To an extraordinary extent, these advances were brought about by the genius of Britons. The world owes cheap steel to Henry Bessemer and Sidney Thompson; accuracy to Joseph Whitworth and Thomas Merton; the telephone to Alexander Bell and the microphone to Hughes; the wireless coherer to Oliver Lodge and the thermionic valve to Fleming; electric light to Swan and television to Campbell-Swinton and Baird. And if we were not first with the car and the aeroplane we certainly redesigned and improved them. British practicality and persistence come out very clearly, and sometimes what comes out is our failure to follow through.

What sort of men were they who made these advances? Let us look at one or two.

Electricity arrives: electrical exhibition at the Crystal Palace, 1892

Charles Parsons
1854–1931

The "strange and weird young man" who revolutionised marine propulsion and made the large-scale generation of electric power possible – among other things.

Like many another famous inventor, Parsons sought for better ways of doing things in many spheres. He invented a highly accurate method of cutting the reduction gears needed by his turbines, known as the Parsons creep. He was interested in optics and invented new ways of annealing glass and of grinding lenses, finally buying up a failing firm of telescope-makers and turning it into the builders of some of the world's major telescopes. For his children he invented toys never previously seen in a nursery, among them miniature locomotives, and *The Spider*, a steam-powered three-wheeler which careered round the lawn pursued by Parsons, his children and three barking dogs.

He would repair his own car in emergencies, though his efforts to mend the kitchen boiler were less successful. He even tried his hand at making strawberry jam. Cook was much impressed. Unlike Henry Bessemer, who had to struggle for an education and again to find an opening for his ideas, Parsons was born with a silver spanner in his mouth. The youngest son of the Earl of Rosse, a noted astronomer, he had access to a workshop from his early years, and received a private education from tutors of ability, before going to two universities. While still at Cambridge, where he rowed in his college crew, he showed his fellow-oarsmen a turbine he had designed and made of paper, wire and sealing wax. "I have an engine which is going to run twenty times faster than any known today" he told them. But they bundled him and his engine under the table.

After a spell with an engineering firm, where he designed his first high-speed engine, and an interlude in which he tried to design a rocket-propelled torpedo, he joined the engineering firm of Clarke, Chapman who asked him to design a marine lighting set. This was his chance. The engineering problem was that conventional steam-engines ran much too slowly to drive the newly invented dynamos, while turbines ran much too fast. Parsons decided to go for turbines; he realised this would involve not only a special dynamo but also bearings capable of standing up to terrific speeds. After tests, at up to 40,000 r.p.m., he settled on 18,000 r.p.m. – fifty times faster than usual – designed a dynamo which would stand up to such speeds, and patented both in 1884. His first plant delivered 7.5 kW.

Four years later he set up on his own, but as Clarke, Chapman owned the patents he had to buy them back. One of the first turbo-alternators he built was a 350-kW set for Manchester Square in London. By 1923, his firm was making units delivering 50,000 kW. He now turned his attention to marine propulsion. No doubt the many trips made in his father's yacht had given him a feeling for things nautical.

Parsons clearly appreciated that his dream-boat, which he named *Turbinia*, would travel much faster than any vessel known and call for a new design of hull. He built 2-foot models and trailed them about a pond at the end of a fishing line. He built a 6-foot model propelled by twisted rubber. But when he tried out the full-scale vessel, the speed was under

20 knots. What was the problem? Measurements of the torque generated by the 2000-h.p. engine showed the power was there. It must be the propeller. In fact, a rapidly revolving propeller leaves an air-space behind the blades (this is known as cavitation) for the water cannot follow them, but this was not known. Parsons built a test tank in which propellers could be "stopped" to all appearances by a flashing stroboscopic light, and discovered what was amiss. He replaced the engine with a new version driving three shafts with three propellers on each; nine in all. In trials the *Turbinia* now achieved the unparalleled speed of $34\frac{1}{2}$ knots. Parsons took her to the Naval Review at Spithead, where she amazed everyone by making rings round the ponderously moving British fleet.

The Admiralty, of course, was hesitant. It would only order a turbine-engined destroyer if Parsons' firm would put down £100,000 guarantee that it would achieve 30 knots. (That would be more than £1m. in today's money.)

H.M.S. *Viper*, as she was named, met the requirements, but was lost in a fog.

Parsons now turned his attention to gearing, realising that the high speed of turbines would have to be reduced. He was able to provide a turbine for a rolling-mill, in which the rollers had to turn at a mere 70 r.p.m. But the gears of the day were not cut accurately enough, and consequently were noisy, and wore rapidly. It was at this point that he invented his cunning gear-cutting machine with a new standard of accuracy. It had the effect of making turbines applicable to any purpose from centrifugal pumps to ocean liners. In 1907, the turbine-engined *Mauretania*, equipped with turbines of 70,000 h.p., won the blue riband of the Atlantic at a speed of over 26 knots.

Though generally remembered for his turbine, Parsons' work on gearing and electrical generation was of greater significance. "His greatness," says a biographer, "stems from his overall conception of the power systems to which he set his hand."

The *Turbinia* at speed, 1894

"A Pretty Scientific Toy"

Although the power of steam was being applied increasingly effectively, a new form of energy was becoming practical which was to change the pattern of life in ways no one foresaw: electricity. By a

A street in Newcastle lit by Swan lamps, 1881

Joseph Wilson Swan invented the electric light bulb

fortunate conjunction, Parsons patented his turbine just at the moment the new dynamos were in need of a convenient source of power. And Swan beat Edison to the post. The first applications were in the field of electric lighting. A few streets and factories were being lit by arc-lights, but these were in need of constant adjustment and the glare was blinding. The idea of a filament lamp had been around for years. Joseph Swan among others had experimented 30 years earlier, but it was impossible to get a decent vacuum. It was the new vacuum pump of 1865, as improved by William Crookes, which really made the incandescent lamp practicable. Swan returned to the problem and showed his new lamp to the Newcastle Chemical Society in December 1878. Across the Atlantic, Edison was experimenting with filaments of carbonised bamboo. By 1881 the House of Commons and the Gaiety theatre were lit by Swan's new lamps; an ocean-going ship, the *City of Richmond*, was equipped and the first electrically-lit train ran from London to Brighton. Electric light had arrived!

Swan believed his ideas were not patentable; Edison patented everything he could. A legal dispute between their firms was settled by amalgamation in 1883. Hence "Ediswan".

Americans, it must be confessed, played a large part in the development of electric lighting in Britain, for many public authorities were convinced it would prove too costly to be practical.

The first large power-station in Britain, at Holborn, was designed by Edison, while Westinghouse, another American, set up a firm to make Parsons' turbines for this purpose at King's Cross. British engineers, sent to America for training, were dubbed the "Holy Forty".

At first people thought in terms of a separate generator for each establishment. The man who saw that you could build central power-stations serving whole districts was Sebastian de Ferranti. The power-station he built at Deptford was intended to serve the whole of north London and was the first to be conceived on this scale. But local authorities elsewhere lacked the courage to follow and Ferranti went back to making equipment. A brilliant inventor, he took out 176 patents in 45 years.

Electricity began also to be applied to transport. Britain's first electric train ran in 1882; the first electric trams were seen in Blackpool in 1886. In industry, electricity was applied not only to drive machinery but to extract metals electrically: zinc, nickel, magnesium and aluminium. This brought aluminium, particularly, down in price to a point where it could be widely used. Electricity also cheapened calcium carbide and carborundum. By 1910 Britain boasted her first all electric coal-mine. The age of electricity had arrived.

A convenient supply of electricity was also a necessary condition for the development of a public telephone service. The telephone can be claimed as a British invention, for Bell was born in Scotland, though his family moved to Canada.

The first telephone exchange was set up in 1879, a rather rapid development considering that Alexander Graham Bell only patented his apparatus in 1876, demonstrating it to the British Association in 1877. A better microphone was invented by Hughes. But the National Telephone Company dragged its feet and the system was taken over by the G.P.O. in 1912.

The Horseless Carriage

As the century drew to its close, a third form of power began to emerge, the internal combustion engine. Dependent on a lighter fuel, and not burdened by a boiler full of water, it was lighter and more compact, thus opening up the possibility of road and air transport. It first appeared in the form of the diesel engine. Despite the German name, the first diesels were British. Dr Diesel was neither the first nor the most important of many inventors connected with oil engines. The first firm to make a successful oil engine was Priestman Bros of Hull in 1885. It was an instant success, starting easily and requiring little attention. Boats and even a locomotive powered by Priestman engines were built. Herbert Stuart improved the design and a pumping-station made to his patents, which started work at Fenny Stratford in 1892 (the year of Dr Diesel's patent), was still functioning in 1939. Subsequently the convenience of electric motors displaced the diesel for many industrial purposes, although their low fuel consumption has retained a niche for them in the field of transport.

The petrol engine, though similar to the diesel, operates at a higher speed. It is usually conceded that the Germans were the first to build a car, despite the fact that Edward Butler's "Petrol-cycle" – a sort of motorised bath-chair – appeared in 1883.

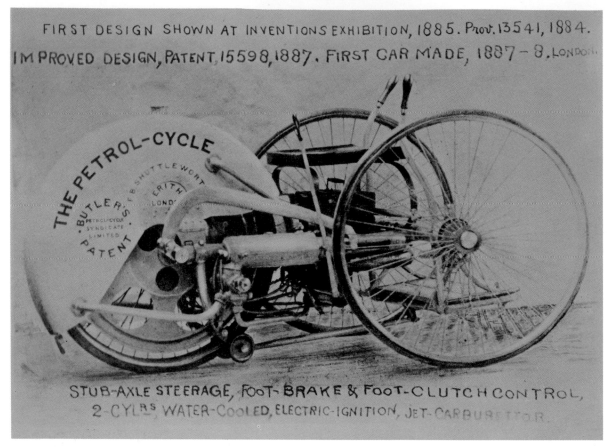

FIRST DESIGN SHOWN AT INVENTIONS EXHIBITION, 1885. Prov. 13541, 1884.
IMPROVED DESIGN, PATENT, 15598, 1887. FIRST CAR MADE, 1887 – 8. LONDON.

THE PETROL-CYCLE
BUTLER'S F.B. SHUTTLEWORTH
PETROL-CYCLE SYNDICATE LIMITED
PATENT
ERITH LONDON

STUB-AXLE STEERAGE, FOOT-BRAKE & FOOT-CLUTCH CONTROL,
2-CYLRS WATER-COOLED, ELECTRIC-IGNITION, JET-CARBURETTOR.

It aroused so little interest that he lost heart. The German Gottlieb Daimler patented his first petrol engine in 1885 and it was about then that Otto Benz produced *his* powered tricycle. In 1887 he actually sold one. He did not produce a petrol-driven four-wheeled vehicle until 1893, the year in which Maybach devised the float-carburettor – but here too Butler had anticipated him.

Britain enters the story as a serious competitor only with the work of Lanchester, to which we come in the next section. The Red Flag law, limiting speed to 4 m.p.h., with a man walking in front, blocked progress. We did, however, make one major contribution to the dawning motor age, with the pneumatic tyre. Actually the first tyre filled with air under pressure, and known as the "Aerial Wheel", was patented as early as 1845 by a young civil engineer named Robert Thompson. Dunlop, a Scottish vet living in Ireland, knew nothing of it when he invented the "pudding tyre" for his son's tricycle in 1888. C. K. Welch of Tottenham completed the job by thinking up the wired edge

and well-rim which made the tyre easy to remove.

The bicycle, of course, was a British idea; and James Starley's safety bicycle with a chain drive to the rear wheel – the first to achieve commercial success – arrived in 1886. (The chain itself was a British invention of 1864.)

Butler's Petrol-cycle, the first petrol-driven vehicle

A chain-driven bicycle of 1878

William Thomson, Lord Kelvin
1824–1907

A combination of Aristotle, Archimedes and Democritus he wrote more than seven hundred scientific papers and became the first scientist to be given a peerage.

If there is just one British scientist in the last hundred years who unquestionably deserves the title of genius, it is William Thomson, later Lord Kelvin. In the whole nineteenth century only James Clerk-Maxwell (another Scot) stands higher but he falls outside our period; he died in 1879.

It is difficult to give an adequate impression of Thomson's achievements for he wrote over seven hundred scientific papers – twelve of them before he graduated – and took out over seventy patents, as well as founding the firm of Kelvin and Hughes and making a considerable fortune. He was the first scientist to be given a peerage. Modern science and engineering are, to a great extent, built on his work.

When he received from London University one of the first honorary degrees it ever bestowed, in 1903, the citation said: "He has promoted true ideas concerning the conservation of energy, the laws of thermodynamics and their application to the universe. He has enquired into problems interesting to the chemist and geolgist and important to the physicist and engineer. He has calculated the probable size of atoms; he has studied the structure of crystals; he has estimated the age of the earth. [Quite incorrectly, as it turned out.] The world knows him best as the man who has shown practically how to measure electric and magnetic qualities. A greater philosopher than Democritus, in him are united the qualities of Archimedes and Aristotle." Brave words!

Grandson of a small farmer of Covenanting ancestry, who had settled in Ireland, his father began life as a labourer and was largely self-taught. But he was appointed a professor soon after graduating at Glasgow University and achieved a national reputation as a mathematician. On his sons he "imposed an exceptionally severe discipline even by nineteenth-century standards". William and his elder brother James received daily in-

Lord Kelvin, drawn by "Spy" in 1897

struction from their father during a pre-breakfast walk at the ages of 3 and 6 respectively. After dinner they studied geography and astronomy. William achieved "a fair competence" in these subjects by the age of 6. His mother died when he was that age and he began to walk in his sleep. He was much attached to his father who let the boys sleep in his bedroom until the sleepwalking ceased. Greek, Latin and French were part of the boys' curriculum. Caesar's *Commentaries* afforded light reading. When the two boys matriculated, James was 12, William 10. At 15, William wrote his first scientific paper. (On Fourier's *Theory of Heat*, which he read in French and mastered in a fortnight, while learning German at Frankfurt.) Admitted to Peterhouse College, Cambridge, he wrote scientific papers instead of concentrating on preparing for his examinations; they had to be published anonymously. A paper he wrote at the age of 17 was thought to be so extraordinary that it was not only published by the *Cambridge Mathematical Journal* but also incorporated into the textbooks. One of his examiners said to a colleague: "We are not fit to mend Thomson's pens."

The University of Glasgow unanimously elected him to a professorship just after his 22nd birthday, despite the fear of some that his lectures might be over the heads of his students. He never prepared his lectures – he was in fact a bad lecturer – he always began them with a prayer. Sometimes he made a discovery in the course of a lecture. He attained popular recognition for his work in connection with the transatlantic cable, for which he had to standardise units and calculate the dimensions, as well as inventing machinery to pay it out. In the course of this he fought many battles with a retired doctor named Whitehouse, who was on the board of the company. Whitehouse had the cable and instruments constructed according to his own amateur theories, and they failed to work until Thomson's instruments were substituted.

But Thomson's greatest achievements were in theory – although he failed to follow up his ideas. He provided the clues which enabled Clerk-Maxwell to develop his electromagnetic theory for light, and Hertz his theory of radio waves, as well as providing a basis for the theory of heat. Despite his interest in the magnetic compass and in gyroscopes he did not see that the compass of the future would be gyroscopic.

This failure to follow through may have been due to his insistence on the social importance of science. "I see no use in a new discovery," he remarked, "unless it is applied to the use of mankind." If he had spent less time on things like the transatlantic cable he might have been the century's greatest figure in the field of science, a second Newton.

His major weakness was inability to listen to the ideas of others: so numerous were his own ideas he would interrupt their explanations. He rejected Darwin's theory of natural selection as incapable of explaining evolution, and could not believe that life arose from a fortuitous concourse of atoms.

In Thomson, or Kelvin as he was to be known, we see the combination of Scottish industriousness with Irish brio and temperament. He could dictate to three secretaries at once, sleep at will; and would immerse himself in calculations at the dinner table or when entertaining distinguished guests, such as the German scientist Von Helmholtz, who was put out by such boorishness.

He believed in a wider extension of university education, at a time when this was by no means accepted. But, in the view of his biographer J. G. Crowther, his real contribution to social progress was to demonstrate to the new governing class the value of applied science.

Lord Kelvin and his compass, photographed in 1902

Rags to Riches

Although Britain was in the van in mechanical engineering, she lagged behind in chemistry, which was virtually a German science. Many German chemists who had been working in England returned to their own country in the period 1864–1880; a few like Alfred Mond and John Brunner (whose firm later became Imperial Chemicals) developed physical chemistry, giving England cheap soap and soda, and making most of the explosives used by Britain in the first world war. The formation by Mond and others of the Society for Chemical Industry in 1881 led to the founding of England's first central research laboratory.

Britain, as everyone knows, was into textiles; Arkwright and Hargreaves had invented spinning machinery almost a century earlier and power-weaving was spreading from cotton to silk and wool in the nineteenth century, and then to the knitting of stockings. Yorkshire "rose on rags to riches". Thus it was not surprising, perhaps, that the first artificial threads should have been a British invention, though it came about curiously. In 1883 Joseph Swan squirted nitro-cellulose through a small hole, in his attempts to make a better filament for his incandescent bulbs – and then with the lateral thinking of genius, saw that what would make a filament would also make a fabric. In the Inventors' Exhibition of 1885, his wife displayed crochet mats made from the new fibres. Two consultant chemists, Cross and Bevan, now came to Swan's aid. By 1892 they had produced viscose (later known as rayon), but when they tried to weave it, the threads broke. As was later realised, the threads had to be left to "age". Courtaulds stepped in and within twenty years Britain was producing over a quarter of the world's rayon.

Sir Norman Lockyer, as secretary of a government commission to enquire into the situation, wrote in 1877: "It is somewhat singular that this country, foremost as it has always been in matters of engineering enterprise, should be so behindhand in the systematic education of its engineers, there being no establishment in England devoted to that object which is recognised by the profession."

Industry was just beginning to see that it was worth backing technical education. Joseph Whitworth set an example by offering thirty scholarships for engineering students. New university colleges developed from the Mechanics' Insititutes and medical schools of industrial towns, their growth accelerated by local pride and industrial munificence. The City livery companies also chipped in, Nottingham, Reading, Exeter, Birmingham evolved into universities.

Despite the reluctance of the Treasury, which repeatedly found excuses for withholding funds, a School of Science (1881), which became a Royal College and finally the Imperial College of Science and Technology (1907), was created.

The World on Celluloid

It was in the 1870s that photography, which for forty years had been a slow and messy business pursued only by enthusiasts, suddenly began to blossom into an amateur activity of vast proportions – almost entirely as a result of British efforts.

Back in the 1830s Fox Talbot had devised the essential feature of a negative from which many positives could be made. The fuzzy pictures of his

Courtaulds' artificial silk was replacing cotton and wool by the 1920s

33

THE DAILY GRAPHIC

An Illustrated Newspaper.

No. 1.   LONDON, SATURDAY, JANUARY 4, 1890.   [PRICE ONE PENNY.

The Weather.

"DULL AND DAMP."

PROSPECTS FOR TO-DAY.

Districts.		Forecasts.
0. Scotland,	N.	Southerly winds, fresh or strong; rain at times.
1. Do.	E.	South-easterly and southerly winds, light or moderate; fair, but damp, to rainy.
2. England,	N.E.	Same as No. 1.
3. Do.	E.	
4. Midland Counties,		Same as No. 1.
5. England (London & Channel).	S.	South-easterly winds, moderate; dull and damp, perhaps rain later.
6. Scotland, (and S.)	W.	Southerly winds, fresh; rain at times. Wind shifting north-west and later.
7. England, (and N. Wales,)	N.W.	
8. England, (and S. Wales,)	S.W.	Same as No. 6.
9. Ireland,	N.W.	Very variable breezes; dull, rainy, unsettled.
10. Do.	S.	Same as No. 9.

THE FOREST GATE FIRE.

A VISIT TO THE SCENE.

(BY OUR ARTIST COMMISSIONER.)

Even so late as Thursday evening a striking testimony was borne to the extent to which the calamity at Forest Gate is felt by individuals not directly connected with it. A large crowd of people, mostly of the poorer class, lingered long after the gates had been closed to interrogate and sympathise with those relations and friends of the children who from time to time issued from the building, and even the policemen at the gates were subjected to constant and eager inquiry. The presence of the crowd was the more striking because from the entrance gates there is so little to see; for the wards which were burnt form so slight a portion of the whole of the buildings as to be almost invisible from that point of view. The bodies, after the fire, were removed to an outbuilding, which is itself a ward. At one end of a passage in this out-

building is a square room, whose schoolroom-like walls are gaily decorated with coloured pictures and prints, such as are dear to the hearts of children; and on the floor, lightly covered with blankets, are the little lifeless bodies. Their faces, in nearly every instance are discoloured; but, except in one sad case, where the arms are raised above the head, as in defence, there is no other visible evidence of the nature of their death, and we may happily believe that at least death was painless. So quiet and peaceful does one little fellow look, that, with the flickering gaslight and the illusion, one can fancy him asleep. And the contrast to this sad scene is that, at the other end of the passage, there is a corresponding room, with pictures on its walls, too, and children there, but alive, and happy and merry. Everywhere, indeed, there is this sad incongruity; and even in the entrance-hall, where, during the last few days, so many grief-stricken parents have waited to identify the victims, there still remain the mottoes with their good wishes of happiness for the New Year.

The Coroner's inquest was held on Thursday, in the Girls' Schoolroom, and the jury adjourned thence to the mortuary, so that there was no repetition of the painful scenes which attended the identification of the bodies by the parents; but, nevertheless, the sobs of some of the mothers, as their children's names were called out, rendered the occasion very saddening, formal as the inquiries were.

Of the origin of the fire there is no certain information, but it is supposed that it occurred from the over-heating of a stove-pipe, which ran from the basement of the building in which the fire took place, through the two dormitories above. This building is separate from the main structure, and is of three storeys. The first contains a needle-room and a wardrobe-room, in which the stove was situated; the second a dormitory, in which forty boys slept; and the third another dormitory, in which fifty-four boys slept. It seems that the first signs of fire were discovered by Miss Bloomfield, who slept on the ground floor of the building, shortly after twelve o'clock, when the bell would be ringing in the New Year. She awoke her colleague, Miss Terry, who slept in the room above, and Miss Terry in turn alarmed the whole house. The doors of the dormitories were locked, and in this circumstance lay the great danger of the situation. Mr. Duncan, the superintendent, was soon on the spot, and, with an extinctor on his back, succeeded in forcing his way to the door of the lower dormitory before he was overpowered by the thick smoke and fell downstairs. Mr. Duncan's heroic efforts to reach the children were seconded by the men connected with the building, and by the firemen who drove up soon after the alarm had been given. The suffocating density of

the smoke, however, made it impossible to penetrate into the dormitory very far, and, to the unfortunate children in the remote corners of the room, no help could be carried. When the flames were subdued—and this was effected in a comparatively short time—it was found that 14 out of the 40 sleepers had perished. In the upper dormitory the tale of death, though heavy, was not so grave as that in the lower dormitory, only 12 out of 54 boys losing their lives. That this was so is due largely to the presence of mind of George Hare, a young man of twenty-two, who slept in a room adjoining the dormitory. Both rooms were already full of smoke when he was awakened by the cries of the boys. Turning to the main staircase, which leads to the dining-room, he found the smoke so dense that he was obliged to shut the door again between the dormitory and the stairs. But there was another staircase at the opposite end of the room leading into the yard, and to this he immediately ran, driving some boys before him, and calling to others to follow him. Happily the door at the bottom of the stairs had already been burst open, and the boys whom he had succeeded in taking with him thus found themselves beyond all danger at once. Groping his way back to the dormitory, he carried and led more of the boys into safety, and when all had thus been brought out, quick and dead together, the energetic young man ran to render assistance in another part of the building. At one end of the upper dormitory two of the female servants slept. Both suddenly awaking to their peril, and seeing no help at hand, one of them with daring readiness climbed down the water-spout, while the other in desperation jumped from the window, being fortunate enough only to break her ankle. The boys themselves do not seem to have been awakened until the fire had made some progress, and even then they were slow to realise their extreme peril. Upon some the smoke seems to have exercised a stupefying effect, and, as far as can be gathered from the narratives of survivors, not only did they make no effort to save themselves, but actually declined to be saved. Many exhibited a curious anxiety about their belongings, one lad being smothered by the smoke while putting on his socks. Another lad of 12 succeeded in rescuing not only himself, but also his clothes basket which stood beneath his bed. Several of the boys, whose ages range between 6 and 12, showed great courage and disregard of their personal safety in helping their weaker comrades.

By three o'clock in the morning the fire was well in hand. All danger of its spreading was at an end, and then arose the terrible question, how many were missing? In the middle of the night,

WHERE THE FIRE BROKE OUT.

with the New Year but three hours old, the still terrified and excited lads were gathered together as well as might be, and the roll was called. Of the 64 names pronounced by Mr. Duncan, only 38 brought the customary response. Of those missing, two were found alive and well in the morning, but the remaining 24 would never answer to any roll call again on earth.

The news of the disaster, like all evil tidings, spread rapidly, and early in the morning many inquiries were made as to the names of the victims. From the fact that the inmates of the schools come from Whitechapel and Poplar, it was impossible that their nearest relatives should come so early, but by noon the doors of the institution were besieged by a pathetic crowd of anxious parents and friends, eagerly inquiring after those dearest to them. Far too many there were whose fears were confirmed, painful uncertainty being only changed for still more painful certainty; but if anything could relieve the terrible impressiveness of the scene, it was the joyful faces of those mothers who learnt that their children were spared to them. But a still more painful scene had yet to be enacted when the bodies came to be identified. The flames and the smoke had so charred and disfigured the bodies of some who had failed to escape from the fatal dormitories that it was impossible to distinguish them further than by a number; this circumstance adding another burden of grief to those who had already more than they could well bear.

It should be mentioned that great assistance was rendered to the firemen and schoolroom officials in the suppression of the fire by some men employed on the Great Eastern Railway, who happened to be passing when the alarm was given. That every effort was made to lessen the fatal character of the disaster evidence from all quarters unanimously shows. Mr. Duncan who himself did so much in rescuing the lads who slept in the lower dormitory, has had the burden of the terrible calamity, which one in his responsible position must specially feel, lightened by a pile of letters congratulating and thanking him for his conduct in the hour of peril. The closing act of this tragedy—the first event which the New Year has brought us—will take place on Monday at the West Ham Cemetery, when the bodies of the twenty-six children whose names are contained in the list of the dead will be interred together.

THE BURNT-OUT DORMITORY.

THE FIRST ALARM.

THE ROLL CALL.

home at Lacock Abbey, now a museum of photography, took thirty minutes to expose. He used paper plates. Then for some thirty years, the wet collodion process was used: you had to pour the emulsion over a glass plate. But Richard Maddox invented the gelatine emulsion, and by 1880 the change over to dry plates was in full swing.

But glass plates were heavy and breakable. Celluloid, invented by Britain's Alexander Parkes in 1861, and then known as Parkesine, was the answer. But it was an American (who changed the name to Celluloid) who first thought of using it as a photographic base. But Celluloid was not flexible enough to be used for roll-film and it was an American company which found a suitable formula and captured the world market.

One offshoot of photography was the fully illustrated daily newspaper: the first issue of the *Daily Graphic* appeared in January 1890.

The French pipped the British to the post on cinematography; the first public showing was held in Paris in 1895, although Edison, who invented the 35-mm perforated film, had shown moving pictures to single viewers through an eyepiece the previous year. In Britain, movies were first seen in 1896. The British, however, were the first with colour cinematography.

A survey such as this necessarily omits much, and overlooks men whose genius was rather specialised. There was for instance, Osborne Reynolds, an expert on liquid flow, who built a scale model of the Mersey and showed that the river's characteristics could be reproduced in miniature. Or again there was Francis Galton, a true genius who not only invented fingerprinting but, more important, launched the science of statistics, which was then developed by Karl Pearson. Darwin was still at work in this epoch. Caton was making his first experiments on electrical stimulation of the brain.

The Healing Art

Medicine, in particular, I have neglected. Sir Almoth Wright suggested a typhus vaccine, two years before the outbreak of the Boer War, in which conflict the typhoid bacillus killed more British soldiers than the bullets of the Boers. Though the use of vaccination had been authorised by the army, it was voluntary and administered haphazardly.

The army decided vaccination was useless, and ordered that it cease. In 1911, the U.S. Army made it compulsory and showed it could cut the typhoid rate to zero.

Wright, though born in Ireland, was professor of experimental pathology at London University, and made other important contributions: his discovery of opsonins was a fundamental contribution to immunology.

Another medical genius was certainly Joseph Lister, who had introduced a crude antiseptic surgery in 1866. Here also was a prophet long without honour in his own land. Bitterly opposed by Simpson, the pioneer of anaesthetics, he continued his work; his fame spread all over the Continent, but at London University's King's College, his lectures were ignored by students, who knew examiners would fail them if they mentioned antisepsis. When the International Medical Congress met in London in 1881, the auditors were told: "England may be proud that it was one of her sons whose name is indissolubly bound up with this greatest advance that surgery has ever made." Belatedly, Lister was made President of the Royal Society, elevated to the peerage and made an O.M.

Also of note were such figures as the Welshman Griffith Evans, who spotted the first trypansome as a disease agent; Sir Michael Foster, whose book on physiology became a standard work; Sir William McEwen, the first man to diagnose and remove a brain tumour; and Alfred King, who in 1882 suggested that malaria was caused by a mosquito. But the palm must go to Ronald Ross who was shown the parasites in the blood of a patient in Charing Cross Hospital in 1894 by Sir Patrick Manson, "the father of tropical medicine", and by 1898 had demonstrated how transmission of the disease occurred.

Ross is a good man with whom to close this section, for he was a man of great versatility, a first-rate mathematician who contributed important articles to mathematical journals, a talented musician who composed chamber music, a novelist whose stories have some merit, and a poet whose poems were praised by John Masefield. "He also invented a system of shorthand and devised a method of phonetic spelling in which some of his poems are written." (1902)

If Ross was not a genius, it is hard to say who was.

Dawn of a New Era
1902–1914

If you were asked what was the most important experiment in modern history, and perhaps in the history of man, you would no doubt reply that it was the experiment which Ernest Rutherford and Frederick Soddy performed in 1902, when they showed that transmutation of the elements was possible and that it released vast amounts of energy. At the time, of course, though scientists were startled to the verge of incredulity by the facts of transmutation, the full implications were not realised, and, to his dying day, Rutherford maintained that the energy locked in the atom had no possible practical application.

The opening years of the twentieth century saw the dawning of a new era in science, one which was to lead to applications as diverse as the computer, television, the electron microscope and the atomic bomb. Though others soon climbed on the bandwagon, the breakthrough took place in Britain. The initial step in this sequence was the founding of the Cavendish Laboratory at Cambridge in 1870, at the expense of the Duke of Devonshire, who was Chancellor. At that time, the idea of a university laboratory devoted to physical experiment was still new; in most places professors did their own research and any instruction students got from them was unofficial. In 1884, when Lord Rayleigh

retired, a new head was appointed, the 28-year-old Joseph John Thomson, the son of a publisher. University greybeards were shocked at the appointment of a mere boy, but he was to make the laboratory the most famous in the world.

Secrets of the Atom

What triggered the wrenching of its secrets from the atom was the discovery of X-rays in 1895. While the rest of the scientific world was using X-rays to probe the bodies of the sick for lost objects and broken bones, Thomson turned them on to a gas and showed that they caused it to conduct electricity, thus opening up a whole range of new experiments. In 1895, following a change in the scholarship rules, workers from other countries began to make a bee-line for the Cavendish, among them the New Zealander Ernest Rutherford. Between 1896 and 1900 these workers published over a hundred scientific papers: a renaissance had taken place in British physics.

Thomson found that the cathode rays (which Sir William Crookes had identified as particles, not rays, a few years before) were deflected by a magnet more strongly than he expected, which suggested the particles were lighter than atoms. After further experiment he announced to a meeting of the Royal Institution that there were particles smaller than atoms – electrons as we call them. Few of those present realised the significance of the report, and some thought it was a leg-pull. How can there be anything smaller than an atom, they asked.

Thomson now performed his most famous experiment. It took days of continous pumping with the primitive equipment available to maintain the necessary vacuum. With a tube, which was the forerunner of the television and oscilloscope tube of today, he balanced the effect of electric and magnetic fields on the particles, enabling him to calculate the mass and charge of the electron. These experiments were checked in the cloud chamber invented by H. A. Wilson, still used in a more sophisticated form today. The passage of a particle leaves a trail of water droplets, the number and size of which can be calculated.

Now radioactivity, discovered by the Curies, enters the story. They isolated radium in 1902 and visited London in 1903. "I would advise England," said Marie Curie later, "to watch Dr Rutherford; his work in radioactivity has surprised me greatly."

J. J. Thomson at the age of 76, drawn by Monnington

It was Rutherford who had already shown that the radiations from uranium were of at least two kinds, and that one of them must be the corpuscles, or electrons, which J. J. Thomson had identified. In 1898, Rutherford took up a post at Montreal and became interested in the radioactive gas emitted by thorium. Here, with a young Oxford chemist, Frederick Soddy, he set to work to analyse it. They had only one hundred-thousandth of a cubic centimetre of gas to work on. They showed that both thorium and its "emanation" gave off particles of immense energy, travelling at 10,000 miles a second, thus explaining the heat generated by radioactive substances. Rutherford perceived that thorium atoms must *disintegrate spontaneously*, hurling out particles and turning into a different element. When he told his Canadian colleagues of this revolutionary idea, some urged him not to publish it, lest it bring discredit on the university. But publish they did, in September 1902. One British professor, incredulous, asked if the atoms had been seized with suicidal mania. Even the great Lord Kelvin could not stomach the strange idea.

I have no space to pursue this extraordinary story (there is a vivid account in A. E. E. Mackenzie's *The Major Achievements of Science*), but one more of Rutherford's achievements must be mentioned: the "scattering experiment". He assigned to two co-workers, Geiger and Marsden, the task of firing alpha particles at gold foil. Most of the particles went through, but a few bounced back. This was, Rutherford felt, "as incredible as if you had fired a 15-inch shell at a sheet of tissue paper and it came back and hit you".

Just before Christmas in 1910 he told Geiger that "he now knew what the atom looked like and what the strong scattering signified". There must be a massive nucleus, from which the projected particles were bouncing, and round it a cloud of electrons, which they could pass through. So Thomson's atom, which *he* saw as a pudding stuffed with electrons, became the nucleated atom we know today.

Such an account, simplified as it is, does scant justice to Rutherford and Thomson's co-workers. There was Moseley – destined to die at Gallipoli, aged 28, with his recall papers in his pocket – who in 1913 used X-rays to determine the electric charge on atoms of different elements and so to show they could be ranged in sequence – hydrogen with one positive charge and one electron revolving round the nucleus, helium with two, lithium with three, and so on. There was Aston who elaborated a method of measuring the masses of atoms. Soddy had found that there were atoms of similar chemical properties but different mass; he called them isotopes. When Aston's apparatus was perfected, he discovered a new isotope every week. This,

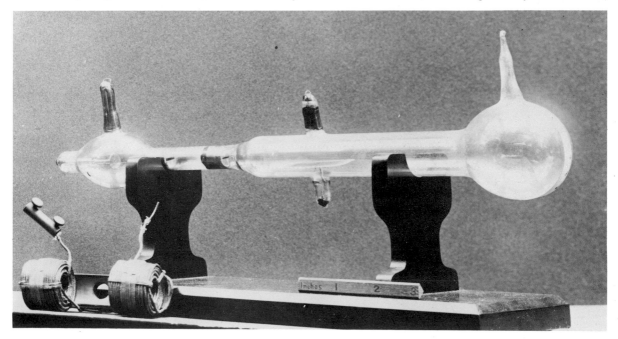

J. J. Thomson's original vacuum tube

however, was in 1919, the year in which Rutherford succeeded Thomson as head of the Cavendish, and belongs to a later chapter in the story.

Now we must go back to 1894, when Oliver Lodge, picking up the early experiments of Hertz who had demonstrated the existence of radio waves in 1888, succeeded in transmitting signals over a distance of 150 yards, and detecting them by means of a Branly coherer (a tube of iron filings). The following year, Rutherford, who at that time was minded to spend his time on Hertzian waves, sent signals across half a mile of Cambridge streets. However, Lord Kelvin told him they would never have a practical application, except to communicate with lightships, and urged him to concentrate on the atom instead. But in 1896 a young Italian named Marconi came to England to take out a patent and Campbell-Swinton recommended him to the Post Office who sponsored his experiments; in 1901 he succeeded in transmitting signals across the Atlantic. The crucial discovery, nevertheless, was Lodge's. He realised that to detect the waves you must have a system which resonates in tune with the transmitter. Marconi only made minor improvements, but such are British patent laws, that you cannot profit from an idea, only from a piece of gadgetry.

Lodge's (and Marconi's) coherer was all right for morse transmission; but to transmit speech something much subtler was needed. This was found in the thermionic valve, devised by Ambrose Fleming, a professor of electrical engineering at London who also contributed much to the development of electric lighting. Combining his knowledge of electric lamps with the new idea of particles being emitted, as in the cathode-ray tube, he saw that the stream of particles could be used as a non-return valve for an electric current. This constituted the breakthrough from which the whole of modern radio and television sprang.

In 1894, just before the advances I have so briefly outlined took place, Lord Salisbury, as President of the British Association, declared: "What the atom of each element is, whether it is a movement, or a thing, or a vortex, or a point having inertia, whether there is any limit to its divisibility, and if so, how that limit is imposed, whether the long list of elements is final, or whether any of them have a common origin, all these questions remain surrounded by a darkness as profound as ever." As

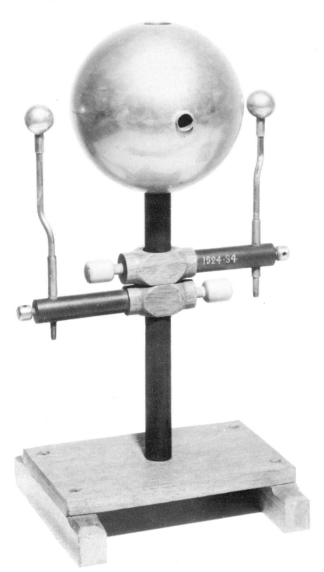

usual, the darkest hour came just before dawn.

The New Biology

The opening years of the twentieth century also saw the birth of biochemistry and indeed the emergence of the new biology which began to probe the mechanisms of living things. It is fair to say that in this great new chapter of science also, Britain led the world.

In Germany a magnificent system of state education was producing a stream of scientists, not only in chemistry and physics, but in psychology and biology too. However, an American biologist observed that Germany produced the generals and the rank and file in the army of biologists; Britain

Lodge's experimental wireless apparatus

produced the commanders-in-chief.

The key figure was Frederick Gowland Hopkins, a modest, patient man who created British biochemistry – until then a German science – from nothing. At a time when many scientists thought living material obeyed different laws from ordinary matter, Hopkins was quite clear that the chemistry of life consisted of "simple substances undergoing comprehensible reactions". But, by an irony of fate, though he devised brilliant technical methods and produced some important findings, in every significant instance he proved to be wrong.

He was not the discoverer of vitamins, as is sometimes said, though he did arrive at the belief that small variations in diet could prove vital. The word "vitamine" (*sic*) was coined by a Pole, Casimir Funk, working at the Lister Institute in London. In 1911 he isolated a substance so potent that it cured beri-beri in fowls in a few hours. It was part of the vitamin-B complex.

While Hopkins was putting biochemistry on its feet, at London University Sir William Bayliss and Ernest Starling were discovering the first hormone and opening up a new understanding of how living organisms are regulated. Twenty years later, Starling was able to say: "When I compare our present knowledge of the workings of the body and our powers of interfering with and controlling these workings for the benefit of humanity, with the ignorance and despairing impotence of my student days, I feel that I have had the good fortune to see the sun rise on a darkened world . . ."

But organisms are controlled not only by chemical messages but also by electrical ones, and here too Britain was in the van. Sir Charles Sherrington, whose description of the brain as "an enchanted loom" has become something of a cliché, was preparing his book, *The Integrative Action of the Nervous System*, which has been called "the greatest landmark in the history of physiology since Harvey's *De Motu Cordis*", the book which explained the circulation of the blood. Sherrington discovered the minute gap between nerve cells known as the synapse.

At this time, too, the subject of heredity was of consuming interest, following the publication of William Bateson's *Materials for the Study of Variation* in 1897.

While biology was thus becoming established as a dynamic rather than a descriptive discipline, the chemists were not inactive. Harry Brearley gave us stainless steel in 1912.

Methods of travel continued to proliferate. The internal combustion engine was applied to tractors, and the first British car, designed by J. H. Knight, rolled down the road in 1895; the first Motor Show was held the following year. The first petrol-driven tractor in the world was seen in 1901, the first Austin car in 1905, while Rolls-Royce produced their Silver Ghost in 1908. It cost all of £985. The first Morris appeared in 1912. In lighter vein, we gave the world Meccano – every boy an engineer – in 1901, while for the ladies, for whom there was already a servant problem, the vacuum cleaner was invented by Hubert Booth, and the Teasmaid device for preparing an automatic cuppa in the following year.

But now the stream of progress was to enter perilous rapids. The first world war was looming.

The "Baby Daisy" vacuum cleaner

Frederick Lanchester
1868–1946

Neglected Genius: "Versatility is not regarded with favour by the British public" said Frederick Lanchester – singer, car manufacturer, aeronautic theorist, engineer and poet.

The only British car entered in the Automobile Club trials of 1899 was the Gold Medal Phaeton, which Frederick Lanchester had produced in the previous year. It achieved the stunning speed of 35 m.p.h. and now rests in London's Science Museum. Thanks to that "Red Flag" legislation which restricted cars to a speed of 4 m.p.h. preceded by a man carrying a red flag, Britain was way behind the Continent in automobile design at this time.

But Lanchester changed all that. Using a small legacy and money raised from friends, he built the first car which was designed as a car and not merely a modified horse-carriage. His car's new features included a rigid frame, dynamically stable steering, epicyclic gears, a new transmission (he had to design a machine to cut the gears), and a magneto generator. He also standardised all the parts, even the bodywork, on all his models – a remarkable thing at the time. But his company, under-capitalised and too much of a one-man show, went bankrupt, despite the fact that it managed to sell cars in considerable numbers. There were no suppliers of components; it had to make everything except the tyres.

Frederick Lanchester was one of the midwives of the twentieth-century world: his discoveries and inventions helped to found the motor and aircraft industries of today. But he never found the right niche. Giving up his studies because they contained no practical engineering (he was doing mining), he invented an accelerometer and a cursor for the slide-rule, but could only get a job as a hack draughtsman at 6d an hour. Becoming assistant works manager of a small firm at a miserable £1 a week, he made some profitable inventions. But aeronautics was his first love. He experimented in his garden with model aeroplanes, and came to the conclusion that vortices form at the wing tips. It proved to be the starting point of modern aeronautical theory, but when he offered a paper explaining his theory to the Physical Society in 1897 it was rejected.

Realising that to build a plane one would need an extremely light engine, he proposed to design one, but his friend Dugald Clerk warned him: "If you were to put forward such a proposition seriously you would be regarded as a crazy inventor and your reputation as a sane engineer would be ruined." He

Frederick W. Lanchester in 1933

therefore turned his attention to the car. Typically, he began by building a motor-boat in order to raise the money to build cars.

After the failure of the car-building venture, he turned back to his beloved aeronautics. In 1907 and 1908 he published two great treatises under the title *Aerial Flight*, one revealing the importance of skin-friction, the other showing how to achieve stability. But his ideas were scoffed at – mainly because of opposition at the National Physical Laboratory, which refused to permit the research work which would have shown their validity. Not until a German mathematician, Ludwig Prandtl, explored the topic after 1920 did acceptance come; the Lanchester–Prandtl theory of lift is now classic.

Lanchester foresaw the great role aircraft would come to play in warfare, and wrote a book about the best way to use them, which was a forerunner of modern operational research.

This many-sided man was musically inclined, and at one point actually was trained as a singer. Characteristically, he invented an improved control for the pianola. He designed his own house and was also an enthusiastic small-boat sailor. Finding himself out of work at the age of 60, he founded a company to make better loudspeakers and quality radio components, but ill-health struck him down. While convalescing, he wrote a long metrical ballad called *The Centenarian*. But observing that "versatility is not regarded with favour by the British public", he published it under a pseudonym.

On his death the Royal Society paid tribute to him in these words: "In him we had one of the very rare combinations of a great scientist, a great engineer, a mathematician, an inventor and a true artist in mechanical design . . . He was also a poet and a philosopher." Since versatility is not liked, it is no surprise that at the end of his life he could barely make ends meet. As his biographer has said: "The grand old man of the motor-car and aeronautics could not himself afford a car or an air fare."

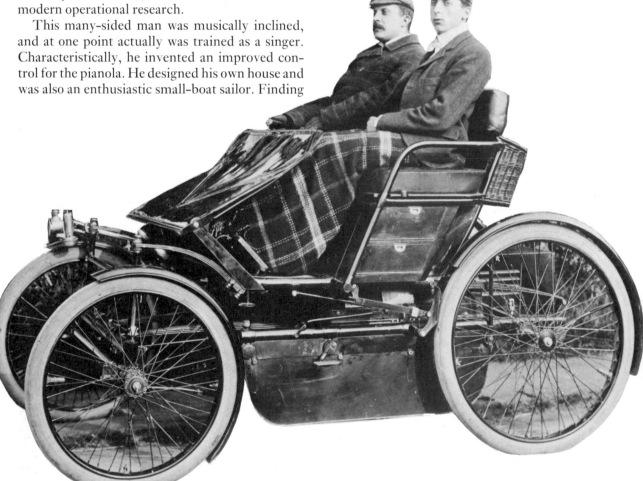

Lanchester in 1898 drives his brother George in his Gold Medal Phaeton

The end of an era: Henley
Regatta on 1 July 1914

The World at War
1914–1918

When the first world war broke out in August 1914, the authorities had no conception of the role that science and technology might play in warfare. The young men who might have become the innovators volunteered or were called up; the older men, with a few exceptions, carried on with their work as best they could.

One of the exceptions was Professor R. W. Boyle, who worked with the Royal Navy to develop the Asdic method of detecting submarines by bouncing sound-waves off them. Though the idea had first been suggested by a Briton, L. F. Richardson, in 1911, the French were the first to conduct experiments. Boyle at once took up the idea, and by the end of the war it was in use. Today, by request of the Americans, it is known as sonar.

What won the war, if any single thing did, was the tank, an idea of which Lord Kitchener remarked: "A pretty mechanical toy . . . the war will never be won by such machines." In its opening stages the war was fluid, which accelerated the development of the armoured car (and incidentally of bullet-proof tyres). When the war bogged down in mud, inventors turned to caterpillar tracks, which had been devised for American farm-machinery. In July 1915, the Armoured Car Division of the Royal Naval Air Service mounted the body and gun-turret of an armoured car on an American caterpillar-tractor. Thus the tank was a naval invention! Independently, Lt-Col E. D. Swinton, an historian, conceived a similar idea: his "Little Willie" (the nickname was a reference to the German Crown Prince) was the first tank designed as such. The man who saw the potentialities of this invention, which the War Office had rejected, was Winston Churchill, then in charge of the Admiralty. His "Landships", as he tactfully called them, were, ironically enough, evolved by a naval committee.

But the military strategists failed to realise that they were a form of mobile fire-power and limited their use to troop support. One hundred were built

Building a bridge over a trench
in France, 1917

The Austin AFT3, a 200 h.p. single-seater fighter triplane, 1918

Armoured cars reconnoitring at the 2nd battle of the Somme, 25 August 1918

A heavy tank, Mark I, the first successful tank

and used in 1916. Eventually more were built. If they had been used in a single massive assault, the war might have ended much sooner.

The war, of course, strongly stimulated the development of the aeroplane. The first powered flight in Britain had taken place in 1908: in 1909 Lloyd George was lamenting Britain's failure to take part in developing this new technology, but A. V. Roe produced his first plane that year, while Short won £1,000 for flying one mile. In 1912, the Royal Flying Corps was formed; Britain started the war with 150 "flying machines" and ended it with 22,000. That put us at the top of the league.

When the Germans began to make night raids on Britain in 1915, using the large rigid airships devised by Count Zeppelin, our aircraft and ground defences, aided by the new explosive bullet, made the ploy too dangerous to pursue. I can just remember, as a small child, seeing a Zeppelin shot down – a burning mass – at Potters Bar.

The war also stimulated the development of "wireless". Directional beams played a major part in the Battle of Jutland in 1916, when the British fleet, under Jellicoe, met the German High Seas fleet in head-on confrontation. (Churchill said Jellicoe was the only man who could win or lose the war in an afternoon.) Our losses were double those of the enemy, for their armour-piercing shells were superior to ours. But the Germans never challenged us again.

But while the war stimulated technological development it gave little scope to genius, except the British genius for improvisation, and for maintaining cheerfulness in appalling trench conditions. Soldiers stood in water-logged trenches, cold and often hungry. Foot troubles were endemic. Spells

First world war recruiting poster

Women shed their skirts: a land girl on a horse-drawn roller of anxious and desperately uncomfortable waiting were broken by massacres. We lost 60,000 men in one day – 1 July 1916.

One activity which could still be maintained, at least in the early days, was the writing of poetry. Names like Rupert Brooke and Wilfred Owen have become part of our literary heritage, although they died so young.

The war brought about great changes in the status of women, who shortened their skirts and flocked into industry and agriculture to take the place of the men who had gone to fight. So it was no coincidence that the vote, for which the Suffragettes had fought in vain for so long, was granted to women over 30. (But it was not for another ten years that it was granted to younger women.) The first woman Member of Parliament was Lady Astor who took her seat in 1919.

As one looks back over the period, oddly enough the name which finally emerges as the candidate for the title of genius is not a general, not a politician, not an inventor or a scientist, but a junior officer, Colonel T. E. Lawrence. A man who could unify the Arab desert tribes on our behalf, winning their respect by his courage and endurance, as much as by his tactical skill, certainly had about him something in the best British tradition. But when you add that he produced a literary work of lasting importance in *The Seven Pillars of Wisdom*, and won the friendship of such undoubted geniuses as Bernard Shaw, only to reject all that fame offered him and mortify himself under a pseudonym as an aircraftman, then it is beyond doubt that we are thinking of a man so unlike his fellows as to deserve, without qualification, the not entirely complimentary label of "genius".

A wireless set of the Army Signal Service, in a dug-out on the Western Front

Colonel T. E. Lawrence at Akaba

Interwar Years
1919–1939

It is difficult today to understand the impact which the Great War had. A whole generation was destroyed. Britain alone lost a million dead, and many more wounded – twice as many casualties as in the second world war. There were no "reserved occupations". Despite the cynicisms of "Oh, what a lovely War!" the casualty rate among the officers was higher than among private soldiers. Not only industrial manpower, but brainpower and leadership were poured down the drain of battle. Perhaps this explains why, in the decade which followed, inventive genius was at a low ebb and a spirit of devil-may-care ruled. It also left hundreds of thousands of young women without hope of marriage.

Of adventurousness there was plenty. British pilots in British planes were first to cross the

The Flying Scotsman, the first
locomotive to exceed 100 m.p.h.
on 30 November 1934

The first Morris-Oxford car,
1913

Atlantic, first to fly to Australia and to the Cape. British cars – handled by Segrave, Campbell, Eyston – repeatedly broke the world speed record. British boats gained the water-speed record; British trains the record on rails. Britons attempted Mount Everest and other peaks. Air travel began to develop: the first passenger service (to Paris in an open two-seater) started in 1919. If there were geniuses, perhaps one was Geoffrey de Havilland whose invention of the slotted wing, making slower landing speeds possible, is still in universal use. Manufacturers began to mass-produce motor-cars: Morris Motors was launched in 1919, the first Austin Seven appeared in 1921. People began to move out of cities, and to travel more freely; the whole pattern of life was changed.

Equally far-reaching was the popularisation of "the wireless" and its offspring, television. I can just remember listening, as a boy, to the barely distinguishable transmissions from Writtle, where Marconi had set up a transmitter in 1920. Two years later the British Broadcasting Company was formed, to be nationalised five years after, when its success was evident. Placed in the hands of a dour Scot, John Reith, a standard was set which has won the grudging admiration of the world.

But what was happening in the scientific field? Edgar Adrian seized on the amplifiers developed for wartime radio and used them to study the electrical impulses in nerves, thus launching electro-physiology. Edward Appleton used radio to study the reflecting layers in the atmosphere which redirect radio waves back to earth. Heaviside had postulated the layers' existence twenty years before; Appleton measured their height and found there were two: they are now known as the Appleton and Heaviside layers. Appleton showed a flash of genius when he realised that here was material for a life's work. Americans had shown some interest at first but did not think it worth pursuing. Appleton discovered how the layers were produced by the earth's magnetic field and, after the second world war brought in a wealth of global observations, produced an embracing theory.

Signs of something rather new appeared in 1926, when John Logie Baird – an inventor so eccentric as to attract publicity – put television on the map. As far back as 1908 Campbell-Swinton had, with remarkable prescience, pointed out that the only way to produce a satisfactory picture would be by means of the cathode-ray tube; and it was indeed with such a tube that high-definition pictures were broadcast in 1936. But Baird, unwilling to wait for the development of a high-definition system, produced his crude stamp-sized pink-and-black pictures and stunned the world. A colour system was demonstrated the following year. Looking back, what does strike one about the postwar decade was the really astonishing number of novelists and playwrights of great distinction who emerged; it became a flood which has never subsided. To these could be added non-fiction writers, such as Julian Huxley or Trevelyan, whose history of England remains a classic.

As a new generation grew up, Britain began to regain vitality. Dirac, aged only 26, predicted the existence of anti-particles, and hence, by implication, of anti-matter – perhaps the most astonishing idea of our time. Two years later, in 1930, he published his great work on quantum physics, for which he was to win a Nobel prize. Two years after that, Chadwick discovered that the atom contained a third particle, the neutron, while, also at the Cavendish Laboratory, Cockcroft and Walton were conducting the experiments on the disintegration of matter which were to lead to the atom bomb. By this time, too, the new wave of biologists was beginning to reach its crest: Bernal, who first applied crystallography to living material; Waddington and Needham, who advanced embryology; Dale and many others. In medicine, too,

Dame Nellie Melba, the Australian prima donna, broadcasting a recital of songs on

15 June 1920: this was the first British advertised programme of broadcast entertainment

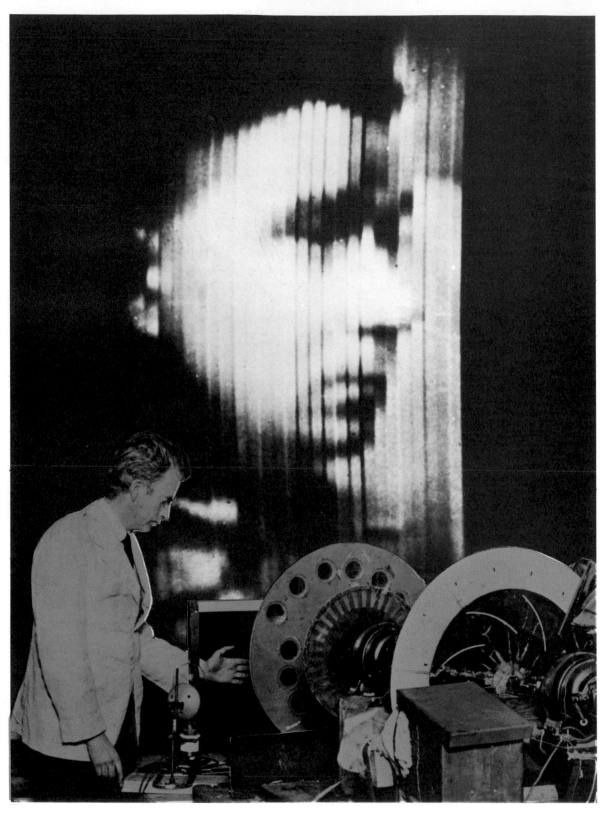

John Logie Baird shows his disc
televisor in 1925

An early Baird picture, 1929,
here enlarged

there was progress; for instance, Andrewes isolated the influenza virus.

Meanwhile in the technological field, Whittle was conducting his first experiments on the jet engine. Initially there was little interest and the first jet-plane to fly was German. At Slough, Watson-Watt was developing an idea which was to prove even more important and far-reaching: radio-location, later renamed radar at the request of America. Perhaps the term "genius" is also deserved by Mitchell who, largely at his own cost, was developing a stressed-skin monoplane, the S6, which was to become the Spitfire which bore the brunt of the Battle of Britain.

Walton, Rutherford and Cockcroft photographed in 1932: Lord Rutherford had already won his Nobel prize in 1908; E. T. S. Walton and Sir John Cockcroft were to share one in 1951

Whittle's jet

The World at War Again 1939–1945

War changes all the rules. Money and manpower cease to be obstacles to invention. New threats trigger off desperate attempts to find a defence. Never was this truer than in 1939, when the second world war broke out. In sharp contrast with the first world war, in the second genius was given its head.

In the earliest days the Germans scattered their magnetic mines which would explode at the mere presence of a metal hull. British scientists responded with "degaussing" – ships were fitted with

St Paul's remained unscathed in the Blitz

Spitfires drove off the Luftwaffe in 1940

huge magnetic coils which neutralised the vessels' magnetic field. Smaller ships were "wiped" much as we now wipe a magnetic tape. Then acoustic mines were dropped. We exploded them by hammering iron plates with pneumatic drills. (Civilians wondered where all the road-drills had got to.) German aircraft began to attack cities and airfields at low altitude. Hence the barrage balloon, the steel cables of which could slice through any aircraft rash enough to penetrate the barrage. Shortages provoked other responses. Because aluminium was short, we made an aircraft of plywood bonded by a new and powerful glue, Araldite. When German aircraft began to ride directional beams to find their targets on moonless nights, British electronics men (notably Professor R. V. Jones) found ways of

Barrage balloons over
Buckingham Palace

Freak effects of an air-raid,
8 September 1940

Radar towers on Britain's east
coast, 1940

Radar interception station

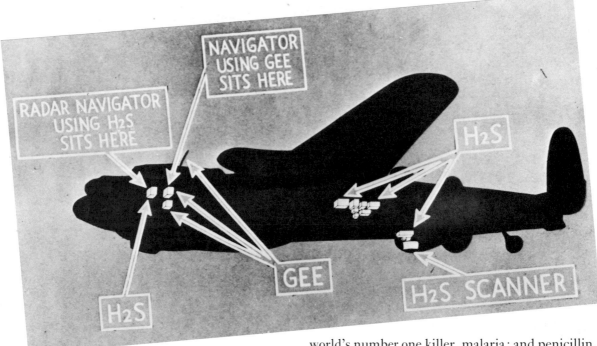

RADAR NAVIGATOR USING H2S SITS HERE

NAVIGATOR USING GEE SITS HERE

H2S

GEE

H2S SCANNER

H2S

"bending the beams". Once this caused the Germans to bomb Dublin, thinking they were over Manchester, an error which did them no good with neutral Eire. (We may be neutral, as one Irishman said, but it's Hitler we're neutral against.)

If the tank was the key invention of the first world war, in the second conflict the key invention was the cavity magnetron, designed on a part-time contract by Professor (now Sir James) Randall and Dr H. T. Boot.

Radar was still feeble and inaccurate. A more powerful source of waves of much shorter wavelength was required. While other workers were pinning their hopes on a device known as the klystron, Randall and Boot reckoned they could pep up a tube, known as the magnetron, by incorporating features of the klystron in it. When it was tried out, it developed such terrific power that it burned out the car headlamps which had been coupled up to provide a rough and ready test. "That day in 1940," says a historian, "can fairly be classed as a turning point in the war." Among other things this device gave us the H_2S radar for bombing through cloud which was the basis of our bombing offensive.

The war was not all destruction, however. It gave us paludrine, an effective weapon against the world's number one killer, malaria; and penicillin, which saved countless lives, developed from Fleming's neglected observation of bactericidal moulds in 1928. Both were British contributions. It also saw the birth of modern chromatography, a delicate means of separating the components of mixtures which has become the universal tool of the biochemist. Based on a Russian idea of 1906, it won a Nobel award for Martin and Synge who updated it.

While the magnetron was the vital piece of hardware, the war also stimulated the development of a fundamental idea: the idea of operational research, by which is meant the scientific appraisal of how to attack a problem. Though the term was coined by Watson-Watt, the radar man, the key name in this field is Pat Blackett, who defined the idea to the Admiralty at the end of 1941. The scientific approach was first applied to the control of anti-aircraft fire, but it soon spread to Coastal Command, where the co-ordination of naval vessels, planes and shore-batteries was made much more effective.

A clear-cut instance: Professor E. J. Williams worked out, during the U-boat campaign of 1942, that if our depth-charges were set to explode at a depth of 25 feet instead of 100 feet, as was the practice, the number of U-boat killings would rise by 250 per cent. The results of making this change

Diagram indicating the position of H_2S radar aids

were so spectacular that the Germans thought we had invented some new weapon. The idea of O.R. spread like wildfire and is now everywhere in use.

The invasion of 1944 produced a rich crop of ingenious devices, including the famous Mulberry Harbours, built in Scotland and towed to Normandy. (If you can't capture a beach, build a precaptured beach and tow it there, someone had suggested.) Trawlers fitted with equipment which gave radar echoes like those of a battleship sailed up the Channel. Model parachutists dropped on miniature parachutes persuaded the Germans we were attacking in the Pas de Calais. Other ingenious deceptions attracted half the German submarine fleet to the coast of Norway.

Meanwhile the jet engine was being developed,

Mulberry Harbour: a
pier giving rapid access
to the Normandy coast

and rockets were being fitted to planes so that they could attack tanks and other tactical targets under ground-control from men on the spot. To convey the multitude of soldiers to France for the first wave of attack, thousands of wooden gliders were constructed and towed behind bombers in long strings, so that they could land in the German rear, while thousands of landing-craft conveyed the main attacking forces. And all the while, under a blanket of secrecy, work was proceeding on the atomic bomb. Cockcroft had had this idea, but in 1940 the prospect of making a bomb seemed remote; it would take a lot of uranium. No one envisaged anything very much more powerful than ordinary explosive. When it appeared that the amount of uranium would not be prohibitive, the task of developing it was handed over to America.

The New Elizabethans
1946–1977

Once again, it took quite a time to shift from war to peace. It is difficult to remember how long it did take for things to return to normal. The street lights did not come on in Piccadilly, nor did clothes rationing end, until 1949, while food rationing lasted until 1954. And, as before, the first achievements represented the application of wartime mechanisms to peaceful ends. Radar was applied to port control in Liverpool; the first gas-turbine car was built; and the first atomic pile in Britain started to function. The effort put into jet aircraft produced the Comet, the world's first commercial jet, and a de Havilland aircraft broke through the "sound barrier". In fields such as these Britain was ahead of the world. We produced the world's first turbo-jet airliner and inaugurated the world's first jet passenger service, from London to Johannesburg, in 1952. In 1958 we started the world's first transatlantic jet service. Today, nearly half the world's jets and turbo-jets are powered by British-made engines. We even sell jet engines to Russia for pumping natural gas. We were also first with vertical take-off: the Flying Bedstead shuddered into the air in 1954; and with the swing-wing, which America took up.

British leadership in exploration was quickly reasserted. Mount Everest was conquered, Kanchenjuga too; the Antarctic was crossed, to name but three achievements.

The Comet takes off on its
inaugural flight to Johannesburg
in 1952

Tensing photographed on the
summit of Everest by Hillary,
29 May 1953

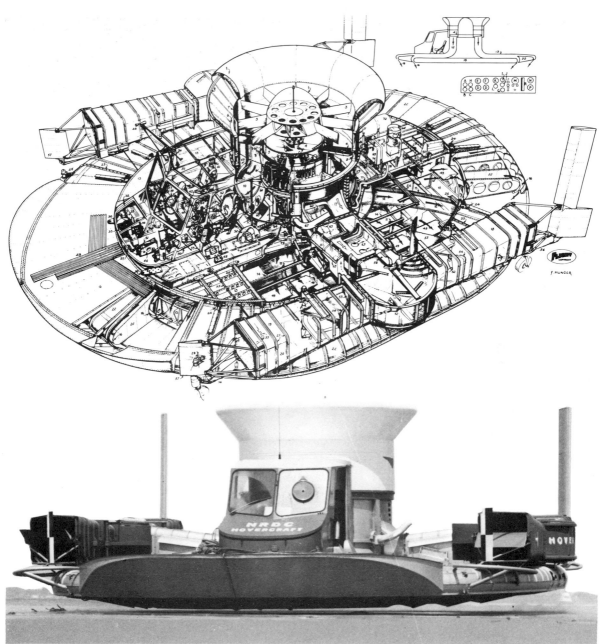

Plan of the hovercraft SRN-1

The SRN-1 on trial

But perhaps the nearest thing to genius in the postwar world was the invention of the hovercraft – or, rather, the whole idea of air-cushion support, which has been called the biggest advance in transport since the wheel. Christopher Cockerell, originally a radio engineer who worked in radar during the war, set up on his own as a boat-builder, but began to think how boats could be built to go faster. Arriving at the air-cushion idea, he demonstrated its feasibility with a hair-dryer and a couple of tins, then spent a year trying to get backing for it. The Admiralty didn't want it: it didn't fit their definition of a boat. Thanks to the backing of the National Research Development Corporation who put £3m. into it, it became a success. SRN-1 underwent its first trials in 1959. Cockerell himself made little from his invention. Air-cushion vehicles have proved invaluable for transporting oil-rigs over Arctic wastes, and played a key role in laying the Alaska pipeline. The principle has also been applied to trains, to supporting badly burned patients, to lawn-mowers, and to moving intensely hot objects in kilns. No doubt more uses will be found.

If we take a long perspective view of the reign of Queen Elizabeth II, we perceive that there were certain areas where we made a considerable effort to maintain our place, others where we bowed out. We sought to keep ahead in aircraft design but, after a promising start, wrote off the challenge of space-flight. We kept ahead for a long while in the field of atomic energy – we were the first country to supply electricity to the public from an atomic reactor – but slowly fell behind here, as also in the field of computers. On the other hand, the government supported medical research handsomely, and British scientists made a great array of brilliant contributions. While falling behind in optical astronomy, for which our cloudy skies make us unfitted, we led the world in radio-astronomy, thanks not to a parsimonious Treasury, but mostly to the public spirit of a manufacturer (Mullards), and a private donor (Lord Nuffield).

It was sad about space. Our Black Knight rocket was launched in 1958 and tested on the Woomera range in Australia. By 1964 we had Blue Streak, intended to be the first stage of a European three-stage rocket. But the European project gradually collapsed in a welter of mutual dissatisfaction. £250m. was wasted. We did however build a series of satellites, put aloft by American rockets, carrying remarkably successful scientific experiments. We have been too modest about this contribution; and few people realise that we are building important parts of the space-lab due to be launched in the 1980s.

The Chemistry of Life
Unquestionably, the glamour field of the 1950s was molecular biology. Crick and Watson's model of the structure of DNA, the molecule which conveys the hereditary message and determines development, released a cataract of experiment which still continues, and has led to the synthesising of a virus and genes, the units of heredity. James Watson was an American, visiting Cambridge unofficially, with time on his hands to natter, while Francis Crick was supposedly working for his higher degree. (When his professor, the famed Sir Lawrence Bragg, learned that he was speculating

about the structure of DNA, he told him to get on with his studies, and "stop rocking the boat".) Subsequently, Crick deciphered the "code" in which the genetic information was embedded. Across the corridor of the prefabricated hut in which this upstart discipline was pursued, Professors Perutz and Kendrew were busy working out the structure of a protein – for reasons of scientific convenience it was the protein which forms the main constituent of blood. Both groups were dependent on the methods of crystallography which Bragg had evolved for the study of non-living materials, though in the case of DNA it was Maurice Wilkins in London who was doing the work. Crick, Watson and Wilkins shared a Nobel prize as did Perutz and Kendrew. Related techniques also revealed the structure of viruses. Professor Bernal, the man who had first applied crystallography to organic material, declared, delightedly: "Life is beginning to cease to be a mystery and becoming practically a cryptogram, a puzzle, a code that can be broken, a working model that sooner or later can be made."

British scientists continued to be leaders in another biological field: the transmission of the nervous impulse and the contraction of muscle under nervous stimulus. Three Nobel awards were given for British work in this field, to Alan Hodgkin and Andrew Huxley, and to Bernard Katz, German-born pupil of A. V. Hill, the Grand Old

Watson and Crick with their
model of the DNA helix

Man of the subject. Also applying physics to biology, another Huxley (Hugh) solved the problem of how muscles manage to contract.

British chemists too were in the forefront, such as Oxford's Rodney Porter who won a Nobel award for his work on the blood substances which confer immunity from disease; George Porter, another Nobel prizewinner, who found a way to study reactions lasting a few millionths of a second; and G. Wilkinson, whose return from a spell in the U.S. put British inorganic chemistry in the front rank.

One cannot mention biochemistry without referring to the work of Frederick Sanger, who decided to try to determine the chemical sub-units making up one of the large biological molecules; he chose insulin. His professor told him the task was impossible; he used to creep back to the laboratory at night to carry on the work secretly. After years of painstaking effort, he produced the result in 1955. In the same year Dorothy Hodgkin, using X-ray crystallography, arrived at a structure for vitamin B_{12}. During the war she had done the same for penicillin.

Since those pioneering efforts, the analysis of molecular structures has become a major field of endeavour in all advanced countries.

The Radio Astronomers

If there is a field of physics of which we in Britain can feel particularly proud, then it must surely be radio-astronomy. Starting from an American wartime observation suggesting that radio "noise" was emitted by the sun, it has emerged that the entire sky is peopled with objects emitting radio pulses, booms and hums (some of them also visible to ordinary telescopes, others not). Thus a wonderful new tool for exploring the universe has become available. British physicists seized the opportunity. Wartime radar had given us steerable "dishes" which could collect faint radio signals from a precisely defined source. Bernard Lovell decided to build, at Jodrell Bank, the world's biggest, 250 feet across – credit that the thing actually worked must go to the engineering firm of H. C. Husband of Sheffield.

When the Americans tried to build an even bigger one, they found themselves unable to solve the engineering problems and abandoned the effort. The big dish also turned out to be just the thing for tracking space satellites: the Russians had to telephone Jodrell Bank from Moscow when they launched Sputnik to find out how it was doing.

As the costs mounted towards a million pounds,

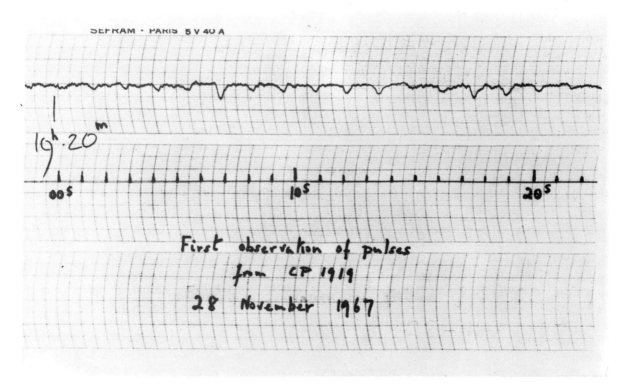

Antony Hewish's first
observation of a pulsar's pulses

far more money than was available, schoolchildren went out to collect sixpences in an attempt to help, and Lovell was told by his university that he faced a gaol sentence for spending money which he had not got. Happily, Lord Nuffield stepped in and saved the situation.

Meanwhile, at Cambridge, Martin Ryle found a cheaper way to achieve much the same result, based on having two or more aerials which are moved about from day to day to cover a large area of ground. Using a computer to combine the readings from each, it gives the effect of an aerial covering the entire area. This proved so successful that an even larger array was constructed, with which Antony Hewish discovered the "pulsars" – radio stars which emit varying pulses of energy. He shared a Nobel prize with Ryle in 1974.

In optical astronomy, the enormous instruments in America and Russia have made the running, but we are now, jointly with the Australians, operating a giant telescope in Australia, and we are constructing an instrument to explore the infra-red part of the spectrum, which will be erected on a mountain-top in Hawaii.

Ryle's work seemed to lend support to the idea that the universe originated in a sort of vast explosion – the Big Bang theory – a view challenged by Professor Fred Hoyle, certainly one of the world's most original thinkers in the field of cosmology, who is maintaining the tradition of Eddington and his predecessors.

If space permitted, I would also boast about our leadership in cryogenics, the study of very low temperatures, and in the related subject of super-conductivity. We have made important contributions to the study of semi-conductors (Josephson won a Nobel award for an insight he had while still a student), and a word must certainly be said about the invention of holography by Hungarian-born Dennis Gabor; if we ever get 3-D television it will probably be thanks to him.

Nuclear Know-How

Britain entered the peace with almost as much atomic know-how as the United States (and certainly more than any third country). We exploded our first atom bomb in 1952, and a hydrogen bomb in 1957, second only to America. We built Calder Hall, primarily as an experimental reactor, but still the first in the world to feed electricity into the public-supply system. We built Windscale, still the world's largest station for reprocessing nuclear fuel. Then things began to go wrong.

After a resounding success with the first generation of nuclear reactors, still producing electricity reliably, we chose for the next phase a design which proved much more difficult to construct than had been imagined, and it has cost twice as much as expected. But at Winfrith Heath, where we built a vast test-site, a third type has been working smoothly since 1967, and could still prove a commercial success. Meanwhile, a fourth variety, the fast breeder, has operated well at Dounreay and a prototype has recently started to produce current. There is public debate about the desirability of employing a reactor which, while freeing us from dependence on South African uranium, produces such dangerous wastes, but technically it looks like a success. Sad to say, we have so far only exported two reactors, for the world decided to wait for something more efficient than our first model – whereas the Americans have sold more than a hundred. Our more recent versions may perhaps do better.

British scientific work in this field has been more impressive. We built the vast Rutherford Laboratory and the £2m. NINA accelerator at Daresbury. Though not of a size to compete with the giant atom-smashers in the U.S.A. and U.S.S.R., it does important work in other areas. We bear the largest share in the costs of the European accelerator near Geneva, however, which competes effectively with Russia and America in exploring the finer structure of matter.

In one area of spin-off from atomic physics we can congratulate ourselves. We have made a business success of radio-isotopes, which have many applications in medicine and in industry, and supply them to many other countries.

Medicine

From this broad field of activity, in which Britain is certainly a leader, it is difficult to know what to pick.

Our surgical successes are almost too numerous to mention. From Russell Brock's hole-in-the-heart operation to the corneal transplants of Professor Batchelor (who runs the world's first eye-bank) our surgeons are a model to the world. And here I must mention the heart-lung machine devised by a young doctor named Denis Melrose in 1949. He has demonstrated it in America and Russia, where a British team showed how to operate on "blue babies". Later Melrose pioneered the deliberate stopping of the heart during surgery.

Heart pacemakers, in which pulses of electric current regulate the beating of a fluttering heart, were developed in Britain and elsewhere about 1960. They suffered from the need for an external electric supply. It was J. Myatt who built the first nuclear-powered pacemaker which could be implanted in the chest. It was fitted to a dog in 1970.

As visitors to the exhibition will see, we have led the world in designing replacements for arthritic joints. The hip joint designed by Dr Charnley of the Wrightington Hospital, Wigan, has been taken up all over the world; since then, replacements for knees and elbows (a much harder job) and for toes and fingers (much easier) have been evolved. We even have tiny artificial "bones" of plastic to replace those within the ear.

Our long experience in virus studies led up to the isolation of the cold virus, which had proved elusive for a quarter of a century, by D. A. J. Tyrrell. It turned out that there were so many varieties that a universal vaccine was not possible. However, Alick Isaacs discovered that the body's cells produce their own anti-viral agent, which he christened interferon. It looked like a big breakthrough, but interferon proved hard to understand and Isaacs died without seeing the defeat of viral diseases for which he had hoped.

More recently Ernest Chain, the man who made the mass production of penicillin possible, has discovered an agent which makes interferon more effective: he calls it statalon. If it is not a virus, it is a very good imitation, and spurs the cell to defend itself. The breakthrough may at last be in sight.

Another British discovery likely to prove of fundamental importance is the existence of substances which control growth – chalones – and so presumably of importance in cancer. This too, may turn out to be the foot-in-the-door which opens up not only cancer control but understanding of how wound-repair and other growth processes operate.

Mention of British medical achievement would be incomplete without a reference to the so-called "test-tube babies" produced by methods devised by Bob Edwards at Cambridge. Seeking to aid women who were unable to procreate, he found how to bring egg and seed together in the test tube

and, after the egg was fertilised and developing, to implant it in the womb of the donor.

Also at Cambridge, but with animal breeding as his practical justification, Rowson discovered how to extract developing eggs from a cow, and insert them in a cow of less distinction to act as a foster mother. In fact, such eggs can be placed in a live rabbit, which is then flown to another country, where the eggs are removed and reimplanted in a local animal. In this way, cows or pigs of one species can produce a litter of another species.

Were there more space, I would say something about our fine record in public health. Hepatitis B, a bane on the Continent, has been kept at bay here.

Bright Ideas

There is an element of genius in inventions of a much more inspirational kind than those we have been discussing. An excellent example is the perforated strips of angled metal from which one can build racks and supports of various kinds. Or again, the letters of the alphabet which can be pressed off a sheet of waxed paper to produce labels,

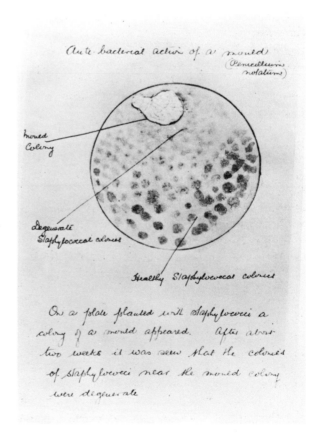

Fleming's notes on his first penicillin growth

headlines for small magazines, and so on. There is an element of genius in the redesign of familiar products in a radical way. The Moulton bicycle was a good example. The Mini car is another; it is not always realised that it is based on a novel form of suspension also devised by Moulton.

One of Britain's commercial success stories has been the development of float glass. The story goes that Alistair Pilkington noticed the grease solidifying on the washing-up water his wife was using (before the day of detergents, obviously) and was struck by the smoothness of the undersurface. It dawned on him that glass could be produced continuously by a process of floating it over a liquid and letting it solidify. It took many years and much money to find the appropriate liquid (it was molten tin) and to iron out the details. According to the great scientist Albert Szent-Györgyi, the discoverer of vitamin C, genius consists in seeing what everyone has seen and thinking what no one has thought. On that definition, we can put Pilkington down as a genius.

Changing World

Already before the war had ended, planning for peace had started. The famous Beveridge Report on Social Security and National Insurance appeared in 1942, while the Ministry of National Insurance was set up in 1944. As soon as a degree of normality was regained, family allowances, free legal aid and other now familiar privileges were introduced. The birth of a new postwar spirit was marked, in 1951, by the Festival of Britain, which expressed a new sense of gaiety and freedom which led in due course to "swinging London" and the mini-skirt.

Reforms and innovations followed thick and fast upon Labour's coming to power. Not only were coal, electricity, gas and transport nationalised, and

Alec Issigonis and the Mini which he designed: unseen, the real novelty of Alex Moulton's hydrolastic suspension

the National Health Service founded, but such reforms as town and country planning, free legal aid and the New Towns Act were instituted. Soon the first of a new series of universities began to arise. Comprehensive education became an official policy in 1964.

As we approach the present day, the details are fresh in memory and there is not need to go over the ground. Key events are the opening of the first stretch of motorway – can it really have been in 1959? – and the introduction of parking meters in the preceding year. We agreed to "go metric" in 1965 after applying to join the Common Market in 1962. Commercial television came in, Centre Point was completed, we won the World Cup at soccer, and the World Cycling Championship. Concorde made its test flight and Rolls-Royce went bankrupt. The government abandoned the idea of a third London Airport, but lowered the voting age to 18 in an unsuccessful bid to stay in power. The Mermaid Theatre opened and the Windmill Theatre closed. It was the era of the Beatles and Carnaby Street, of atomic energy and the mini-skirt, of *Morning Cloud* and Mary Quant.

Sir Stanley Matthews played football for England fifty-six times

Sir Leonard Hutton captained England's cricket team between 1952 and 1954

Roger Bannister breaking the tape to become the first man in the world to beat the four-minute mile on 6 May 1954

Laurence Olivier as King Henry
addressing his soldiers before the
battle of Agincourt in the 1945 film of *Henry V*

Three
Goodbye Moralitee

But the circumstance which mostly encourages the arts in England is the great veneration which is paid them.
Voltaire Letters concerning the English Nation

As my poor father used to say,
In 1863,
Once people start on all this Art,
Goodbye moralitee.
And what my father used to say,
Is good enough for me.
A. P. Herbert Lines for a Worthy Person

"During the last thirty years England and Wales have achieved a richer flowering of the arts of music, opera, drama, literature, painting and sculpture, and a greater increase in the enjoyment of them than in any comparable period since the start of the industrial revolution," declared Lord Redcliffe-Maud in a recent Gulbenkian lecture.

Perhaps there is some connection between the fact that, precisely at the moment when Britain's industrial achievement is deficient, her achievements in the creative arts is at a peak, as all the world acknowledges. There is an antithesis between the grinding perseverance and reliability which develops a new product, manufactures it, maintains quality, sells it and services it, on the one hand, and the spontaneous self-expression which is the basis of creativity. This is not to say that artists and writers do not work hard. The dedication which is necessary to become a ballet-dancer is well known; pianists of brilliance practise for many hours a day; architects labour over their plans; and even authors have to discipline themselves to sit at their typewriters in awful solitude and cudgel their brains. Be this as it may, Britain is on a cultural pinnacle.

I suppose the theatre is the most English of arts. Shakespeare's genius is acknowledged by the admirers of Racine and Goethe, of Calderon and Ibsen, of Chekhov and Strindberg. For a time, Victorian rectitude inhibited the British playwright. The first signs of rejuvenation were seen in the tuneful parodies of Gilbert and Sullivan, never successfully imitated. By the close of the nineteenth

Sir Arthur Wing Pinero, whose most famous play was perhaps *The Second Mrs Tanqueray*

century, the Irish Bernard Shaw and Wilde, and the English Pinero were infusing wit and style into the London theatrical scene, where performers of the standard of Henry Irving, Ellen Terry, and Mrs Campbell were setting new standards of acting. After the intermission of the 1914–18 war, the theatre resumed its position, with the plays of Coward, Lonsdale, Ashley Dukes and others, while designers like Gordon Graig and Rex Whistler, and impresarios like C. B. Cochran and Basil Dean achieved international fame.

The 1930s, when London had more theatres functioning than any city in the world – over fifty at one time – was marked by the rise of new playwrights such as Rattigan and Ustinov, Bridie and Emlyn Williams. Actors of world-renown became so numerous that one cannot list them; hardly a Hollywood film but had at least one British actor in the cast. Names like Charles Laughton, Edith Evans, Lewis Casson, Sybil Thorndike, John Gielgud, Peggy Ashcroft, Laurence Olivier, Constance Collier, Marie Tempest, Gladys Cooper, Leslie Howard, Cedric Hardwicke compete for mention.

In the postwar period, the tide of theatrical genius shows no sign of abating: Osborne's *Look Back in Anger* may be taken as a milestone of the new realism, even cynicism, and the rejection of the middle-class domestic comedy in favour of the kitchen-sink comedies of Wesker and the surrealism of a Pinter, a Stoppard, or a Mortimer. And if Tyrone Guthrie was the world's most sought-after producer of the 1930s, Peter Brook assumed that mantle in the 1950s. The Marat/Sade play in the Charenton lunatic asylum was a theatrical conception for which there were no real precedents.

And how is one to acknowledge the vitality of Joan Littlewood in the East End, of Bernard Miles in the docks at the Mermaid, of Stratford or the English Theatre Company at the Royal Court? What is one to say about R.A.D.A. or the vigorousness of provincial repertory? New theatres of unusual design have been built at Chichester, Birmingham, Manchester, and elsewhere, culminating in the National Theatre of Denys Lasdun, which has been called by far the finest national theatre in the world.

In architecture, Britain has not produced names of the unique character of a Corbusier, a Gropius or a Frank Lloyd Wright; nevertheless, a level of excellence has been maintained. Now that private patrons are a thing of the past, it is public buildings, such as schools, cathedrals, theatres, hospitals and offices which offer the architect an opportunity to shine. The efforts of postwar architects have been geared more closely to social needs in Britain than in any other country, and that is a major achievement.

Sir Henry Irving, one of the last of the great actor-managers

But modern Britain has made one other contribution of permanent value to the visual environment – the evolution of town and country planning. Reacting from the grime and smoke of the industrial slum, it was Ebenezer Howard who conceived the Garden City. Letchworth was built in 1903; Welwyn in 1920. Industralists, too, began to plan: Bournville and Port Sunlight are their legacy.

Dame Margot Fonteyn, Prima
Ballerina of the Royal Ballet

It was no flash in the pan. Today, Britain has a system of town and country planning which, with all its defects, is more advanced than any other country's, and has done much to preserve the landscape. Moreover, it is an exercise in planning democratically. There is a certain antithesis between the idea of democracy and the idea of planning, which implies stopping some people from doing what they want to and forcing compliance on others. With our system of public enquiries, and the interlocking of local and national authority, we may be hammering out a more genuinely democratic administrative technique which will prove a model for the world to follow.

However, a more natural train of ideas from the theatre would be ballet and opera. Before the war, when it was fashionable to be foreign, Peggy Hookham called herself Margot Fonteyn; Lilian Alice Marks took the name Alicia Markova, and Patrick Healey-Kay rechristened himself Anton Dolin. Today, Anthony Dowell and Antoinette Sibley need not pretend to be anything other than British. In this renaissance, one must recognise the contribution of Marie Rambert and Ninette de Valois as well as of such choreographers as Frederick Ashton (who studied under Rambert) and Kenneth MacMillan, or John Cranko.

In opera we cannot compete with Germany, where it attains the status of a national sport, but we can be proud of the world-famous Royal Opera at Covent Garden and the English National Opera, and the unique venture of Glyndebourne, as well as the more modest opera productions of societies and local music clubs. And this is perhaps the place to mention the Edinburgh Festival, and the many festivals which have followed its lead, at Cheltenham, Bath, Aldeburgh, and elsewhere.

In music-making too, we can boast of internationally known names such as Benjamin Britten and Michael Tippett, as well as performers of international rank like the pianist Clifford Curzon and Dennis Brain – before he died, the world's leading horn-player. It was about 1900 that British music began to escape from Victorian inhibition: it was really Edward Elgar who placed British music on the map. His work, like that of Vaughan Williams, Holst and Delius, found a wide public thanks partly to the Promenade Concerts, a unique institution, created by Henry Wood – a man who could fairly be classified as a "genius".

Sir Henry Wood, the conductor
who created the Proms

Deriving his love of vocal music from his Welsh mother, he nevertheless gave the Proms an international character, before dying during the fiftieth season. Once, he so enraged the musicians of the Queen's Hall Orchestra that fifty of them walked out and founded the London Symphony Orchestra, now the most recorded orchestra in the world.

Having mentioned Wood, one is bound to add hastily: Beecham, Sargent, Barbirolli. But it is probably Benjamin Britten who has best exemplified the element of genius in British music. Of him, no less an authority than the Russian composer Shostakovich has said: "I consider Benjamin to be a great composer. Along with Mahler, Prokofiev, Miaskovsky, Stravinsky, Bartok and Berg, Britten can be classified as a classic of the twentieth century." In 1945, the first opera to be produced in any European capital after the war was Britten's *Peter Grimes,* based on a poem by George Crabbe, who had lived in Aldeburgh some two centuries before. Very English it was, with hornpipes, church bells and tunes in pubs, and it enchanted the audience. But it did more: it brought Britain from the ranks of opera-performers to the select company of opera-creators, a position which Michael Tippett and others have helped to maintain.

King Cophetua and the Beggar-Maid (detail), painted by Sir Edward Burne-Jones in 1884

Image 11, sculpted by Dame Barbara Hepworth

Dazzle-Ships in Drydock at Liverpool (detail), painted in 1921 by Edward Wadsworth, a member of the Vorticist group

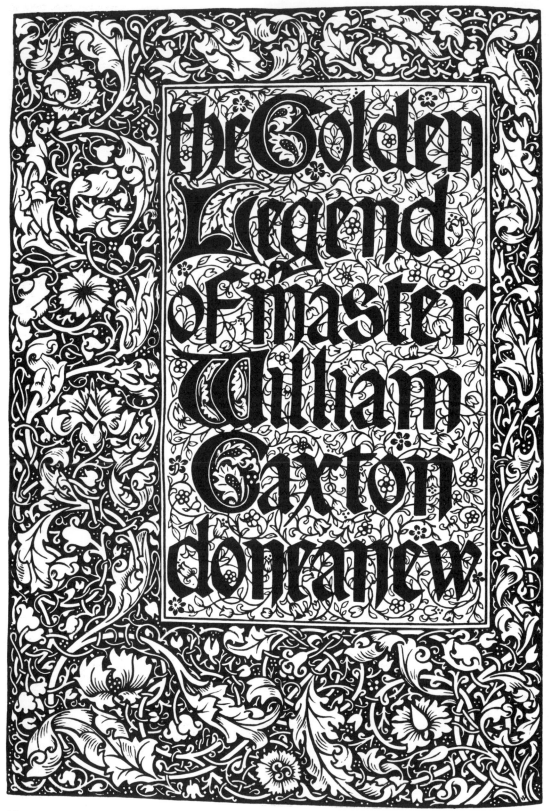

The opening pages of *The Golden Legend* printed at the Kelmscott Press

OF THADUENT OF OUR LORDE.

HE tyme of thaduent or comyng of our lord in to this world is halowed in holy chirche the tyme of iiii wekes in betokenyng of iiii dyuerse comynges. ℂ The i was whan he came and apierid in humayn nature and flessh. The ii is in the herte and conscyence. The iii is at the deth. The iiii is at last Jugement. The last weke may vnnethe be accomplissed. For the glorye of the sayntes whiche shal be yeuen at the last comyng shal neuer ende ne fynysshe. And to this signyfyaunce the first responce of the first weke of aduent hath iiii verse to rekene ℂ Gloria patri & filio for one to the reporte of the iiii wekis, and how be it that there be iiii comynges of our lord, yet the chirche maketh mencion in especial but of tweyne, that is to wete, of that he came in humayne nature to the world, and of that he cometh to the Jugement & dome, as it apperith in thoffyce of the chirche of this tyme. And therfor the fastynges that ben in this tyme, ben of gladnes and of joye in one partie, & that other partie is in bitternesse of herte. By cause of the comynge of our lorde in our nature humayne, they ben of joye and gladnes. And by cause of the comyng at the day of Jugement, they be of bitternes and heuynes.

As towchyng the comyng of our lord in our bodyly flessh, we may considre thre thynges of this comyng. That is to wete thoportunyte, the necessyte & the vtylyte. ℂ The oportunyte of comyng is taken by the reson of the man that first was vanquysshyd in the lawe of nature of the defaulte of the knowledge of god, by whiche he fyll in to euyll errours, & therfore he was constrayned to crye to god ℂ Illumina oculos meos, that is to saye, lord gyue lyght to myn eyen. After cam the lawe of god whiche hath gyuen commandement in which he hath ben overcome of Impuissance, as first he hath cryed ther is non that fulfilleth, but that comandeth. For ther he is only taught but not delyuerd fro synne, ne holpen by grace, and therfore he was constrayned to crye, ther laketh non to comande

I have left myself little room to discuss painting, still less, sculpture. There are certainly far more art galleries than ever before. But how is one to summarise a stream of painting which extends from the pre-Raphaelites to Francis Bacon, Bridget Riley, John Piper, David Hockney? Perhaps one could pick as particularly English the industrial townscapes of L. S. Lowry or the whimsical village scenes of Stanley Spencer. But then one would have omitted Aubrey Beardsley and Burne-Jones at one extreme, and Augustus John and Graham Sutherland at another, the *Yellow Book* and *Blast*! Sculpture is rather easier; here, Henry Moore and Barbara Hepworth tower above their rivals.

Even a survey as brief as this must acknowledge the extraordinary explosion of interest in "arts and crafts" which has taken place in the last few years. Here, too, there is a long tradition, to be sure. William Morris could be called the founder of the movement, back in the 1890s. He battled for well-made, unfussy furniture, elegant wallpapers, good printing. He founded the Kelmscott Press. More than that, he crystallised a vision of Utopia which the British found peculiarly attractive – a world of small communities, good fellowship, simple pleasures, which he described in his famous *News from Nowhere*.

The communes and the alternative society movement of today are direct inheritors of this tradition. But the torch has been carried on by people like Eric Gill (whose typography we still see in every London tube-station), by collectors of folk-songs like Cecil Sharp, by glass-engravers like Laurence Whistler, by the makers of stained-glass windows like Hugh Easton, by potters like Bernard Leach, by designers of fabrics and furniture, by makers of cutlery and silverware, by designers of jewellery and goldsmiths, too numerous to mention.

A German writer once observed that the English have the reputation of being inartistic, and that even many Englishmen think this to be true. But, he pointed out, architecturally England is one of the richest countries in Europe, and this alone should demolish the idea. "One may therefore say," he concluded after making a survey of the subject, "that art in the true sense of the word has got so thoroughly into the bones of large sections of the English people . . . that it is no longer recognised as such." He could even be right.

A chair of ebonised wood, designed by Charles Rennie Mackintosh in 1902–3

Last but certainly not least, we must pay tribute to the achievement of writers – writing is undoubtedly the field where Britain shines most brilliantly. A glance at the bookstalls in New York, Bombay, or any European capital, or a walk round the Frankfurt Book Fair, would at once show that British books outnumber those of any other nation – with a population one-quarter that of the United States, we publish more new titles than they do. But our achievement is not merely in quantity. Whether in fiction or in non-fiction, the names of scores of British authors are internationally known – from novelists like D. H. Lawrence, Evelyn Waugh, Graham Greene, Lawrence Durrell, John Cowper Powys, and thriller-writers like Ian

Fleming, Agatha Christie, Michael Innes, to heavyweights like Arnold Toynbee, Bertrand Russell, and George Orwell, to name but a few – and British writing is supreme. And this is a field where women have particularly come into their own. One thinks of names like Virginia Woolf, Rebecca West, Vita Sackville-West, Ivy Compton-Burnett, Iris Murdoch, Muriel Spark, Margaret Drabble . . . And to be sure we have something of a speciality in children's books: Squirrel Nutkin, Noddy, Dr Doolittle and Winnie the Pooh are now part of the culture.

It is of course a long tradition. A century ago people were devouring the works of Thomas Hardy and Anthony Trollope; then it was H. G. Wells and

Bertrand Russell,
mathematician, philosopher,
writer, pacifist

George Orwell, author of
Animal Farm and *Nineteen
Eighty-Four*

George Gissing. Half a century ago it was Galsworthy and Barrie, Arnold Bennett and Somerset Maugham. And these giants looked back to an even earlier generation; to Jane Austen and George Eliot, to Dickens and Fielding, to Sterne and Defoe.

Was it the remarkable power and flexibility of the English language which made England a nation of writers (even the diarists seem to have been inspired), or was it something in the British character which created the language required for its expression?

Virginia Woolf, novelist, photographed in 1902 when she was 20

Given such a splendid instrument, it is not surprising that England produced such noble word-smiths as Shakespeare and Milton, as the translators of the Bible and the Lake Poets. And it is quite in this tradition that English writers continue to produce new coinages: brillig and slithy, runcible and quark. Surely Lewis Carroll and Edward Lear are uniquely British – and here we must not exclude the Irish who have long excelled in the poetic and inventive use of the Queen's English.

It is all the more surprising that English poetry has been in a long eclipse. In our time perhaps only W. H. Auden comes near the genius level.

Foreigners who speak English with real fluency agree that its resources far outstrip those of French and German, not to mention other tongues. Dr Renier, a Dutch writer critical of the English, as may be judged from the title of his book *The English Are They Human*, has nevertheless called English "a wonderfully balanced and adequate instrument, not only harmonious and full of sonority – for these are not qualities precluded by [English] barbarity – but supremely polished, precise, exacting and definite. In the course of my experience as a translator I have been surprised to find that, unlike the German and the Dutch, the English language is impatient of approximation, of prolixity, and of the shameless torturing of the phrase upon the rack of completeness." He explains this as due to our character. "It is, in short, the language of precise and meticulous observers and reporters of facts, of philosophers who do not dwell in the clouds of metaphysics so much as on the pasturelands of pragmatism, of subtle psychologists who do not circumambulate by devious roads. It is the language of vigorous, clear brains, of mental adults." Coming from a man who has declared that the English "completely differ from human beings", that looks suspiciously like a compliment.

H.G. Wells, whose many books included *The War of the Worlds* and *Kipps*, at the age of 74

W.H. Auden, the poet, in 1938

Sir Noel Coward, actor,
playwright, composer, at the age
of 36

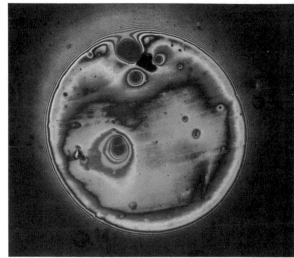

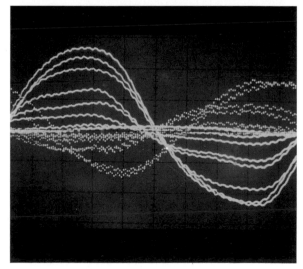

**The changing face of British
ingenuity**
Concorde
vacuum cleaner advertisements
hovercraft

test for tactile variations in rubber
pithead
recording wave-power

Electricity may continue to come
mainly from coal – we have vast
reserves

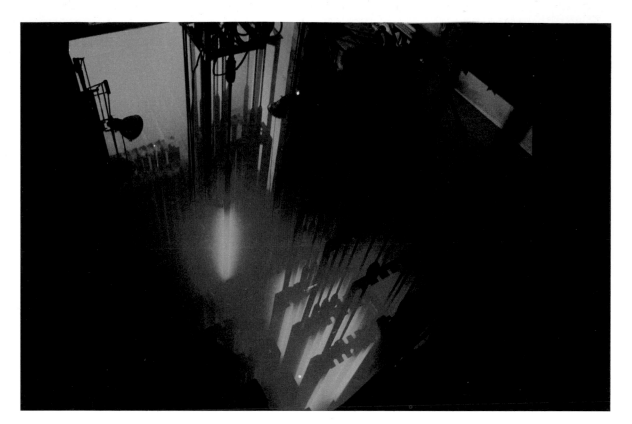

Dirty or clean energy: nuclear
energy or wind-power (here
used to heat greenhouses)?

From light bulb to laser: less
than one hundred years after
Swan's first light bulb, Gabor
used the laser to create 3-D
images

North Sea oil gives us thirty years
to solve our energy problems

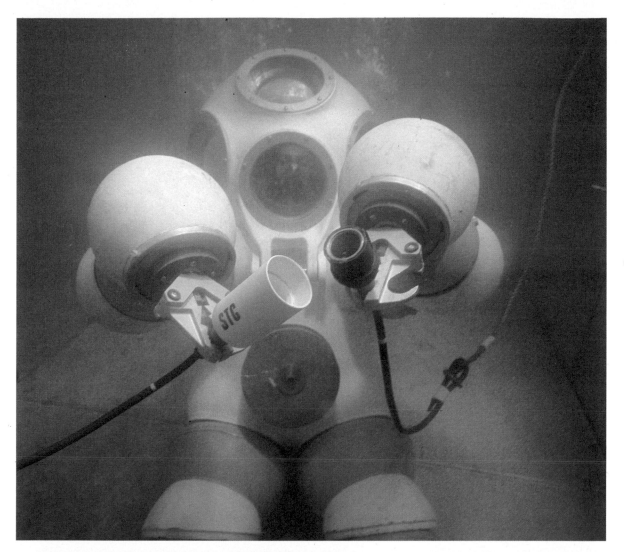

The Jim "shirt-sleeve" diving-
suit protects the deep-sea diver
from cold and pressure and
frees him to perform complex
tasks

Graham Roxborough (right)
stands beside a model of his
Scotbuoy, the floating oil-rig

Lillian Pavey is a prolific
inventor, among whose many
devices is this music-score
typewriter.

Barnes Wallis, famed for the
Wellington bomber and the
bouncing bomb, is still at work
on sub–orbital flight

The Advanced Passenger Train is British Rail's latest contribution to inter-city transport

The Lynx helicopter's solid rotor head has cut cost and maintenance: it reduces the number of working parts from about 140 to 40

The Harrier "jump jet" is a key British contribution to Western defence

Fully automated steel-making:
Britain's largest steel bed at
Rothermere

These Friesian cows gave birth
to Hereford calves, thanks to egg
implantation, a British technique
to aid animal breeders

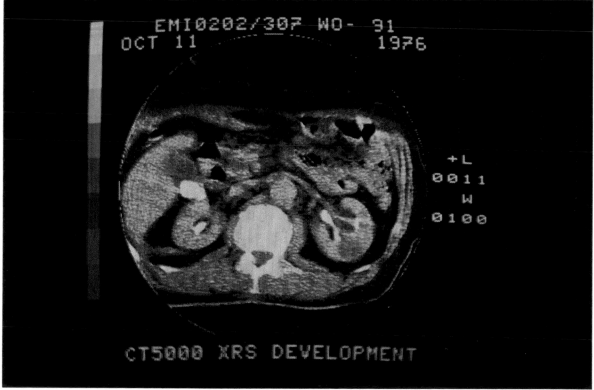

The E.M.I. body scanner
(above) has revolutionised
diagnostic medicine. The "slice"
shows the patient's vertebral
column flanked by his kidneys.

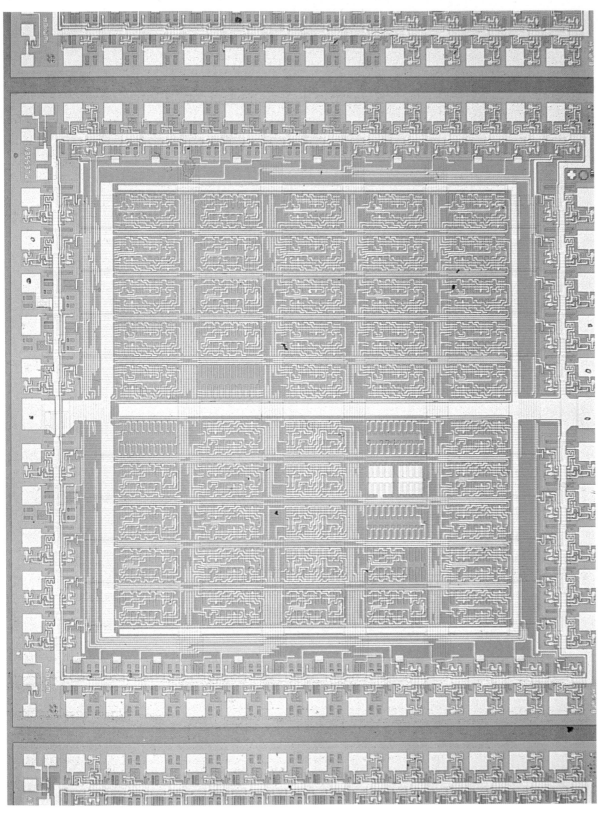

Plessey's computer circuit chip, in which thousands of transistors are integrated, helps to shorten and cheapen the design phase of computing equipment

The electron microscope
introduces man to worlds within
worlds, even an atom: this is a
minute part of a thumbprint

Is the fast breeder reactor –
economical on fuel but
producing highly dangerous
radioactive wastes – a hope for
the future or a white elephant?

Four
The Secret Vigour

So . . . I feel in regard to this aged England . . .
pressed upon by transitions of trade and competing
populations, – I see her not dispirited, not weak,
but well remembering that she has seen dark days
before; – indeed, with a kind of instinct that she
sees a little better in a cloudy day, and that, in a
storm of battle and calamity, she has a secret vigour
and a pulse like a cannon.
Ralph Waldo Emerson, an American 1847

Concorde, the world's first
commercial supersonic
aeroplane

The tremendous advances in science and technology, to which British genius has contributed so much, have created serious problems. Medical advances, especially those which have slashed infant mortality, have created the population explosion. Technology has wolfed resources as never before. In ten years America alone used more raw materials than had been used in the whole of history. Intensive farming methods have, in many places, damaged top-soil and soil fertility. Britain, in addition, faces the problem of inflation and of inability to match exports to imports. The needs of industry, and the breaking up of traditional patterns by travel and education, have created problems of social unrest, violence and crime.

In the British Genius Exhibition can be seen a selection of the new inventions and ingenious ideas which will affect our future and help solve some of these problems. Here a number of contributors provide a background of ideas, discussing the problems to which these exhibits are related.

The five topics to be discussed are: energy, food, technical change, health, the home.

We start with energy, to which all our problems come back eventually. Fortunately, the energy outlook for Britain is unusually favourable, and British inventors are evolving novel ways of improving our position still further.

Challenge No. One
Lighten Our Darkness

In January 1974, Sir Kingsley Dunham, chairman of the Council for Environmental Science and Engineering, together with sixteen prominent British engineers representing a wide range of activities, wrote a letter to *The Times*, saying that the oil-price crisis, serious as it was, was "minor compared with the long-term problem of energy supply which faces the entire world". The signatories considered that the information given to the public about the future availability of fossil fuels was "misleadingly optimistic". The outlook for oil was "most alarming". They argued that the highest priority should be given to developing British fuel resources – an effort was needed comparable with that effort to re-arm at the beginning of the second world war.

In the short run, Britain is cushioned by the discovery of oil and natural gas in the North Sea and in the seas to the west as well. Britain also has enormous resources of coal. These may well sustain her into the next century when advanced reactors and possibly fusion power will be available. It is unlikely therefore that the lights will go out, even if we are required to share our good fortune with our Common Market partners. However, electricity

Part of an oil-rig, dwarfing all around it, transported by road

and gas are no substitute for oil: no one ever devised an electric aeroplane. Oil is also needed as a source of chemical raw materials, and can even be converted to edible protein. Oil will remain costly, unless effective means of extracting it from shale or making it from coal are invented. Until then, to consume it in power-stations which could be fuelled by coal or nuclear energy is folly.

What no one can be sure about is whether demand for energy will continue to expand at the same rate ($3\frac{1}{2}$% p.a.) as in the past: in the essay which follows, Peter Chapman, director of the Energy Research Group at the Open University, argues that it may level off, leaving us with, perhaps, an energy surplus. The Central Electricity Generating Board has not ordered a new power-station for some years, the industrial concerns which make power-stations are in jeopardy. A new station may have to be ordered just to keep them going until they are really needed again. Other experts have predicted a glut of oil by the mid 1980s, following intensive prospecting and more economical use.

Genius is at work seeking ways to extract energy from the environment which do not depend on fossil fuel. Some are experimenting with windmills. The main difficulty is that a windmill light enough to turn in a summer breeze is liable to be wrecked by a gale, even if the windmill-blades are arranged to "feather" over a certain speed. Others are looking to the sea's waves, which could meet Britain's entire electrical needs if suitably harnessed. Others believe energy could be obtained from the tides: a single installation could meet one-tenth of our annual needs – but it would cost £2,000m. Solar cells, as used on satellites, which convert sunlight directly to electricity, would solve the problem effortlessly, if only they were not so costly. The sunlight which falls on an area the size of a tennis court every hour is equivalent to about 5 gallons of petrol. And of course you could save the cost of distribution by making the current just where you want to use it. Experts do not expect solar cells to supply a significant amount of energy before the end of the century – and meticulous engineering will be needed even then.

As the Duke of Edinburgh has urged, we must use the short breathing space which North Sea oil has given us to do more research into novel energy sources with a sense of real urgency.

Taking Energetic Measures

by Peter Chapman

Although the U.K. is on the verge of becoming a net exporter of energy the government is planning to increase coal production, by opening new pits at Selby and possibly elsewhere, and is debating spending £1,000m. on a new type of nuclear reactor. Even though everyone recognises the inflationary effects and hardship caused by increases in fuel prices the government has recently instructed the Gas Corporation to raise its prices and may instruct the Central Electricity Generating Board to purchase more power-stations than it needs, thereby increasing the price of electricity. What are the reasons behind these apparently contradictory decisions? Why is there such a fuss about leaks of radioactive material from Windscale and what would happen if there were another oil embargo by the Arab nations? With so much coal, oil and gas, why are some people getting so anxious about building bigger nuclear power-stations and windmills? And if the real problem is getting energy cheaply enough, why can't we copy the astronauts and get our energy from the sun?

Most of these questions about energy and energy policy revolve around different views of the future prospects for energy supply and demand in the U.K. We can all remember the havoc caused by the miners' strike and the last oil embargo, and no one would like to go through the experience by choice. It is now obvious that our industrial society is as dependent on energy as you and I are on food and drink. The problem is trying to decide just how much we will need in the future and exactly where we will get it from.

Of course "need" is a difficult word in the humdrum of politics. Our grandfathers certainly coped with life with very much less energy than we use today. And even now the French people manage to have an almost identical lifestyle to our own but use exactly half as much energy per person as us. Part of this is due to the warmer climate in the south of France, but the largest differences appear to occur in industry and transport, indicating that the French manage to use their energy a lot more efficiently than we do. But even if we were to use

energy more efficiently we should still need a tremendous lot of it, and, if past trends are a guide to the future, then we will need a lot more in twenty-five or fifty years' time. And here lies the problem, namely that it takes between twenty-five and fifty years to build up a large energy-supply capability in anything except oil and coal and gas. These three fuels have enabled the industrialised nations of the world to grow at a remarkably steady rate over the past twenty-five years. But oil and gas deposits are becoming depleted and new deposits more difficult to find; and coal production is limited by how many of us (yes – us, not them) are prepared to go and work underground. To get behind the present rush of energy problems it is necessary to start by looking at the U.K.'s prospects for energy production.

Planning Ahead

The largest U.K. producer of energy at the moment is the coal industry, but unless it manages to modernise its mines it may soon play a very small role in our energy economy. The problem of the coal industry is that it has inherited many Victorian mines that seventy years later are really uneconomic. To be able to produce coal at a competitive price the National Coal Board will have to

open many new mines like Selby where the amount of coal produced per miner per year is five times the present average. With more mines like Selby the supply of coal could rise to 200 million tons per year by the year 2000; representing between a third and a half of our total demand by then. With present reserves it could also maintain this production for at least a hundred years. Although we presently have a surplus of coal production it is necessary to start investing in the new mines now if we are to have them available in twenty-five years' time.

A similar situation applies to the electricity industry. At the moment we have twice as much generating capacity as we need to meet the maximum, or peak, demand for electricity. But we know that many of the existing power-stations will be obsolete in twenty-five years' time, and, since it takes between seven and ten years to build a power-station, we must start building now for the year 2000. Furthermore if the supplies of oil, coal and gas and uranium are going to be short (or expensive) by the year 2000 then, so the U.K. Atomic Energy Authority argues, we should be building a new type of reactor that "breeds" its own fuel. The problem is that breeder reactors are very expensive (about £1,000m. each) and are potentially very dangerous – certainly a lot more dangerous than the

The coal-miner's "sniffer": a modern methane meter

existing nuclear reactors. Apart from dangers from the reactor itself there are very serious problems associated with the transport and reprocessing of the reactor fuel (plutonium) since this is the material used in atomic weapons. Some observers think that the widespread development of breeder reactors will lead to a proliferation of nuclear weapons with all that that involves.

The oil and gas industries face a much easier situation, although it is likely to be shorter-lived. It only takes a few years to develop an oil or gas field, even in the unkind conditions of the North Sea, so it is fairly easy to increase oil or gas production quickly – provided of course there is some oil or gas in the ground. U.K. production of oil and gas is expected to reach its peak in the early 1980s, when we will have a sizable surplus for export. But from about 1985 onwards the production will be falling off as the fields become exhausted – and we only have one North Sea.

Energy Gap?

Estimated total production of coal, oil and gas for the U.K., on fairly optimistic assumptions, is shown here. Also shown are two estimates of the growth of demand. The high-growth demand shows an "energy gap" appearing in about 1995,

the low-growth curve shows a smaller gap ten years later. Most of the questions raised at the beginning of this section are concerned with trying to predict which of the demand curves will actually apply and how to fill either of the "energy gaps".

The argument advanced by those who forecast an "energy gap by 1990" is based largely on the continuation of past trends. Over the past twenty-five years energy consumption has grown at a steady 2% per year and if this continues up to the year 2000 then our energy demand will be between 600 and 650 m.t.c.e. by then. (In order to lump different forms of energy together, we can state them in terms of "coal equivalent" – the amount of

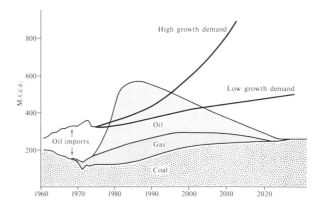

The probable demand for fuel and the available supplies in Britain: we would be in trouble by the end of the century

A 45-ton nuclear fuel flask is delivered to the Oldbury Nuclear Power Station coal you would need to produce the amount of energy in question. Thus "m.t.c.e." means millions of tons of coal equivalent.) Since it is expected that we will only get 150–200 m.t.c.e. from coal and 250–300 m.t.c.e. from oil and gas, there is a "gap" of 100–200 m.t.c.e. between the supply and the demand. The argument for continued growth in energy demand is linked to an assumption about continued economic growth. As economic growth continues then we all become a little better off and are able to afford to buy more things, including appliances which consume fuel. Our extra purchases have to be made by industry which needs to use energy in its production processes. Thus, so the argument goes, there is a strong link between economic growth and growth in energy demand. The argument can also be supported by pointing to countries like Canada and the U.S.A. where, on the average, people are about twice as rich and use about twice as much energy.

Levelling Off?

The argument against this case is based on the assumption that historical trends in energy demand will *not* continue. This is justified by two arguments. Firstly, it is argued, the new higher prices of fuels will encourage people to use it more sparingly. Householders may take steps to insulate their houses and buy cars which do more miles to the gallon. Similarly, industrialists will be looking for ways to cut their energy costs by improving the efficiency of their machines. Perhaps more importantly, it is pointed out that many of the factors that give rise to the historical trends are very unlikely to continue. One of the most significant trends over the past fifty years has been the increase in the number of households, even though the total population has grown fairly slowly. This has come about because the average number of people per household has decreased from 5 per household in 1900 to 2.7 per household in 1975. Whilst a sizable fraction of the population are children this trend cannot continue, so the future growth in the number of households is likely to be closer to the growth in the total population, which is very small. The number of households is important since it is this that determines the number of heating systems, cookers, water-heaters, refrigerators and so on. In other words it is the number of households that determines the number of energy-consuming

and appliance-consuming units in the country.

In addition to the slowing down in the growth in the number of households, there is also likely to be a slowing down in the growth of domestic-energy consumption per household because this will "saturate". For example, whilst everyone was cold there was plenty of scope for growth in energy consumed in heating systems. But now that 50 % of all dwellings are centrally heated the *rate* of growth must slow up. This is illustrated in the second figure which shows a transition from everyone being "cold" to everyone being "comfortably warm". Even if you were wealthy, you would not want to heat your house to 25°C (88°F) because this is too hot to be comfortable. Since about half the houses in the U.K. are centrally heated now, we are at the "turning-point" in this S-shaped curve, and the historical trend (shown dashed) will not continue into the future.

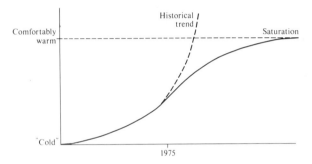

Similar arguments apply to the ownership of motor-cars which passed through the turning-point in 1965. Also a wide range of electrically-operated household appliances are well past the turning-point and approaching their "saturation level". Now if households are not going to buy many more appliances, cars, houses and heating systems, then industry will not need to make many more of these, so it will not need to increase its energy consumption very much. Thus, so the argument goes, neither the domestic nor the transport not the industrial sectors of the economy will increase their energy demand as fast as they did historically. It is argued that we will be spending our extra wealth on things like automatic control systems, videophones, increased services and so on, all of which require much less energy for both their manufacture and for their operation than cars, cookers and heating systems. Advocates of these arguments point to countries like France which

Since the demand for home heating saturates at a comfort level, it will not grow as fast in the future as in the past

enjoys a higher standard of living than the U.K. but with about half the energy consumption. On the basis of these arguments energy demand in the year 2000 is estimated to be between 400 and 450 m.t.c.e. – which is between 50 and 100 m.t.c.e. *less* than the projected energy supply.

The differences between these two projected demands are important because the lower projection gives us an important extra ten years *and* a smaller energy gap to fill. In fact the recent performance of the U.K. economy and the performance of the nuclear industry, which is the only energy source large enough to fill the large energy gap, suggest that if the high growth occurs then we will be in some difficulties. We will not be short of energy to keep warm, at worst we might find that we have to import a lot of expensive energy and run ourselves into the depressed balance of payments that has plagued our economy for the past three years. This will retard our economic growth, reduce the amount of capital available for investment and hence reduce our ability to increase energy supplies. So not only does the high-growth

demand curve seem implausible, it also seems undesirable.

In contrast, the low-growth curve has a lot of attractions. There is a wide range of energy sources that could make a substantial contribution to meeting the energy gap. A modest nuclear

A model of Salter's "duck": the curved structure under the water at the left rocks with the waves to extract energy from them

Professor Stephen Salter

programme, with, say, 20,000 megawatts of re-actors by the year 2000 and twice as many by 2025, would fill half the gap. We could get 20,000 MW of electricity by installing a long string of Stephen Salter's rocking "ducks" off the Atlantic coast of the British Isles. These ducks are large concrete structures which extract energy from the waves very efficiently. This system has the advantage of producing most output in the winter, when we want most, since it is then that the waves are highest. We could get a similar contribution from the sun by paying more attention to the design of our houses. By far the simplest way of capturing the sun's energy is to have a large, unobstructed, south-facing window. If this is provided with shutters to keep the heat in at night, and a blind to prevent summer overheating, it could provide up to 30 % of all the heat needed to keep a house warm. We could get hot water by using the solar "flat-plate" collectors shown in the Exhibition. We might even get some useful contribution from windmills if we could develop a suitable energy store to tide over

The solar cell module converts
sunlight directly into electrical
power

the calm days. Solar cells, which convert sunlight to electricity, would be even better – if we could cut the cost dramatically. The advantage of having so many possibilities is that we do not have to rely on any one technology to a high degree. If Stephen Salter's ducks do not match up to his design calculations, or if nuclear power proves too dangerous, then we still have other sources that we can turn to.

But none of these sources will be available in the year 2000 or even 2025 if we do not start developing them seriously now. Part of the controversy associated with the energy-demand forecasts is connected with "who gets the money" to develop new energy sources. The U.K. has only a certain amount of cash to spend on energy research, and it is crucial that the right choice be made. There is a lot of British Genius at work in developing nuclear systems, wave-machines, wind-machines, solar systems – and not least systems that enable us to use energy more wisely. There is, unfortunately, no genius around who can tell us which is the best bet.

Challenge No. Two
Thought for Food

We can no longer depend on a plentiful supply of cheap imported food. Shortages are becoming general and prices of food are rising steadily. They will almost certainly continue to do so. Essentially, this is a world problem. It is estimated that 70,000 people starve to death every day somewhere in the world. High prices mean prosperity for farmers. But in Britain the problem is complicated by the need to export to balance our imports. We import about half the food we eat; and if allowance is made for the animal-feeds and fertiliser which we import, the figure is nearer three-quarters.

Between 1963 and 1973 the government ran three programmes for expanding home production. But in that time the value of food imports has more than doubled: more was imported and at higher prices. Thanks to climatic changes and rising world populations, prices will rise further.

One solution is to eat less meat. The land which would feed animals to provide meat for 100 people would grow grain or vegetables for 500 or more.

But the problem goes deeper. In Britain there will be 3 million more mouths to feed by the end of the century, while the land available for agriculture steadily shrinks: a million acres has been withdrawn from use in the last twenty-five years. Farmers, too, have been abandoning the traditional method of ley farming (which means resting cropland every few years by putting it under grass,

to restore its natural fertility) with serious consequences: an Agricultural Research Council report revealed that soil-structure is being damaged and pests and disease are increasing. Many small farmers have been taken over by the food-processing industry. Others have had to specialise in the most profitable crops, notably cereals.

The more efficiently we use farm labour, the more we drain the life from the countryside, turning it into a mechanised food-factory for town-dwellers. In a recent survey, two-thirds of those questioned said that they would prefer to live in the country. So what is the sense of a policy which produces just the opposite effect? After all, the final objective is to live the kind of life we like in the kind of society we like. The production of food is only one component in that situation.

In short, the high efficiency of British farming in terms of output per man and output per acre conceals a disquieting underlying situation. An increasing number of people feel that the solution is not to press ahead with more mechanisation, more specialisation, and more intensive methods but to revert to methods which have stood the test of thousands of years, while making sensible use of new strains and methods where appropriate.

So sharp is the split that I decided to present the positive and negative aspects separately. Let us start by looking at the many ingenious advances in farming technology which British genius is evolving. Afterwards, Robin Clarke, a practical farmer, reveals his deep concern about the whole direction farming is taking and suggests changes.

Experiments show the result of controlled crop-fertilisation: the wheat on the right has received no fertiliser

Farming's Third Revolution

British agriculture is the most highly mechanised in the world: there are more tractors than men on British farms. Less than 2 % of the population work in agriculture yet we produce about half the food we eat. Production is also outstanding in yields per acre.

Though the average size of farms has doubled in the past ten years, most of Britain's 164,000 farms are under 300 acres in extent. They remain preponderantly family businesses.

What has made this high productivity possible is not merely the tractor (a British invention, by the way, of 1902) but the numerous machines for milking, reaping, stacking, milling, grinding, crushing and sorting which are also used. And in this field progress remains rapid, much of it led by the Institute of Agricultural Engineering at Silsoe, Bedfordshire. The rotating milk carousel is becoming a standard on large farms: with its aid, one man can milk 85 cows per hour. Even higher figures are foreseen for the future – up to 300 cows per hour.

Fruit-picking is another job calling for tedious hand labour. In the near future we shall see machines like the Hydrapic Twin which will strip a hundred acres of blackcurrants in three weeks – before the crop rots. And the Hydrapic can be rejigged, when picking is over, to serve as an ordinary tractor. Another slow job is sorting potatoes, which arrive mixed with earth and stones. Now X-rays are being used to distinguish the two. Engineers are also working on ways of removing the leaves and stems mechanically. Of every three potatoes grown, only two reach the table, the third being so damaged by the sorting and cleaning processes as to be unacceptable. To find out how

The Hydrapic blackcurrant harvester shakes the berries from the bushes and delivers them into trays, having separated unwanted leaves

this occurs, Dr D. McRae has made an artificial potato containing a radio-transmitter and sensors. It will go through the whole sequence of processes with the regular potatoes, sending out cries of distress when it is being treated unkindly.

Another laborious job is thinning rows of seedlings which have germinated too freely. This back-breaking task can now be lightened by using jets of water to wash out the unwanted seedlings.

Other machines are being devised for sowing seed more effectively. Some seeds, such as celery, have very poor rates of germination; sometimes only 2% of the seed sown actually develops into plants. This wastes land, labour and seed. A field which is 98% bare earth is hardly worth the effort of cultivating. So Dr J. K. A. Bleasdale of the Agricultureal Research Council's Vegetable Research Station at Wellesbourne is engaged in a five-year programme aimed at getting the seeds to germinate first and then to sow only those which are really trying. To sow a germinating seed without breaking off the delicate growing tip is not easy; but a machine which will do this has already been devised. It washes the seeds into a slit in the soil with water. The problem of storage has also been solved. What is proving more difficult is to get the optimum conditions for germination. It may take ten years, and £100,000.

The advantage of tractors, reapers and similar machines, it should be emphasised, is not merely that they save labour but also that they enable the farmer to profit from spells of fine weather in our unreliable climate. If there are only two fine days in a month of rain, and you can get the crop gathered in these two days, you may avoid losing most of the crop. Incidentally, we have a tremendous export business here. In 1975 exports of agricultural machinery were worth £628m., against £419m. the year before.

Research for Farmers

However, devising new machines is probably not the most important, even if it is the most obvious, of the efforts being made to improve our farming. We spend, nowadays, some £50m. a year on agricultural research – more than ever before – which is spread over a wide range of activities, from the design of farm buildings to toxic chemicals, and from animal diseases and plant pests to the study of weeds. Particularly rewarding is the breeding of better strains of plants and animals.

Before the breeders got at it, corn stood shoulder-high. But tall corn is easily knocked flat by rain and then cannot be harvested by reaping

Maris Fundin, the first British
semi-dwarf wheat, is examined
by its breeder, Dr Lupton

machines. Moreover today there is a much reduced demand for straw. So plant-breeders have concentrated on producing dwarf varieties. Much corn now is less than 3 foot high and breeders talk of getting down to 18 inches. Unfortunately, dwarfing corn seems to increase the number of branchlets bearing no ears of corn (known as "tillers") while many of the new varieties seem liable to disease, and crop poorly if weather conditions are less than perfect.

Over-fertilising is almost as harmful as under-fertilising (and of course unnecessarily expensive). Research shows that it is quite common. Studies are being made of the optimum time and rate of fertiliser application. Other studies are aimed at detecting disease in time to apply preventive measures. There is a good prospect of increasing yields of wheat from the present average of less than 2 tons per acre to 4 or even 5 tons.

Much of this work has been done at the Plant Breeding Institute at Cambridge, while the Scottish Plant Breeding Institute has devoted attention to growing wheat and grasses in colder, wetter conditions. The Welsh Plant Breeding Institute at Aberystwyth (which has a sizable export trade in seeds) has recently succeeded in developing a new strain of rye-grass combining high yield with the ability to survive in adverse conditions.

A crop which is difficult to harvest commercially is the edible pea, which sprawls all over the place unless tied to wires or supported. The John Innes Institute at Norwich (named for the renowned inventor of John Innes compost) is developing a pea-plant which has no leaves but many tendrils which cling to one another so that the plants support themselves. But so far these leafless peas do not provide much food: the problem is now to build in a higher protein content.

Apple trees which grow straight up, so that the fruit can be gathered mechanically, are also on the way; the orchard of the future may look very different.

Animal-Breeding

Animal-breeding is also receiving attention. The use of artificial insemination, enabling prize bulls to fertilise thousands of cows without leaving the home farm – even cows in other countries – was pioneered in Britain, at the Animal Physiology Institute at Babraham. The next step was to implant eggs, removed after fertilisation from high-grade cows, in run-of-the-mill cows which thus acted as unwitting foster-mothers. At first this involved operating on the mother-cow, which made it impractical for ordinary farming purposes, though demonstrating the workability of the idea. But now methods of sucking out the fertilised eggs are being evolved, and this may become a widely-used procedure in the near future. This is also British work. (See the colour picture on p. 92.)

But in addition to research on specific problems of the kinds I have mentioned, we are beginning to recognise the need for strategic as opposed to tactical research: how can we best dispose of the country's assets in land, labour, equipment, genetic stocks, and know-how? A bold development, which I think has no parallel elsewhere, is the founding of a new Centre for Agricultural Strategy, under the aegis of Reading University, with the aid of a £250,000 grant from the Nuffield Foundation. Our membership of the European Economic Community has changed the rules of the game: the Continental countries have long preferred to protect farmers rather than the consumers of food. As Britain imports £3,600m.-worth of food a year, this is no laughing matter. We need radical ideas. (That we could combine orchards with public parks is one way-out suggestion, which has been turned down for fear of vandalism.)

Fish-Farming

We have always been – surrounded by the sea as we are – eaters of fish. But overfishing and the incursions of mechanised fishing-fleets into local waters make the outlook discouraging. The price of fish is rising. Many people are therefore turning their thoughts to fish-farming; that is, the deliberate raising of fish in optimum conditions in tanks (properly called mariculture) or in arms of the sea, such as the lochs of Scotland. There are already some 220 fish-farmers in Britain, with their own union, now incorporated in the National Farmers' Union. They produce about 1200 tons of high-grade fish and shellfish a year, worth about £1m. Forecasts of an eventual output of 20,000 tons a year – some optimists speak of 50,000 tons – are seriously made. Some fish prefer warmish water and Britain has recently launched an experiment in fish-culture in waters warmed by the cooling-water from Hunterston Atomic Power Station. The fish

cultivated here are the commercially valuable sole and turbot, and costs are down to £500 a ton.

The use of tanks could be even more effective. The area which would yield 100 pounds of beef could produce a ton of fish or 100 tons of shellfish. The Chinese have been cultivating fish – mostly carp – in ponds for centuries and produce 1½ million tons p.a. in this way. Carp, once a staple Friday food in this country, have gone out of fashion; but experiments are being made by a team from Chelsea College, at Rye Meads, Hertfordshire, in growing carp in water enriched by the wastes from the nearby town of Hoddesdon. They put on weight rapidly.

Unfortunately, fish do not like overcrowded conditions and often die if bruised, so that in all enclosed fish-farming culling is a critical operation. And we know very little about the diseases of fish, which could spread rapidly in crowded conditions.

Such difficulties present a challenge which we can overcome.

Another approach to the growing fish-famine is the use of electric currents to attract fish to points where they can be netted. In the past high voltages have been used, but recent British studies show that quite low voltages tickle plaice and shrimp agreeably. This could ease the work of inshore fishermen – remember that 42% of our fish are still caught by small inshore vessels. Sonar aids for spotting shoals of fish are being developed for such vessels by the White Fish Authority.

Way Out

Further in the future lie possibilities which could change the whole farming picture quite radically.

First, there is a long-standing attempt to transfer to other plants the remarkable capacity of legumes (e.g., beans) to incorporate nitrogen from the air.

Salmon are farmed in this
fish-farm in Lochailort,
Inverness-shire

Other crops have to get their nitrogen in the form of fertiliser – no less than 36 million tons a year are applied, world wide, and many countries are using much less than the optimum. Legumes make their own, which is why we plant clover and plough it in to improve the soil for corn and other crops. Legumes do the trick with the aid of specialised bacteria, known as rhizobia. If we could persuade these bacteria to inhabit the roots of non-leguminous plants the whole world fertiliser position would be changed. It would be the farming breakthrough of the century. Workers elsewhere have shown the rhizobia *will* coexist happily with non-leguminous plants, in special circumstances. So how to make a habit of it?

At Sussex University there is now an A.R.C. unit of Nitrogen Fixation, where a multidisciplinary team is studying the problem in all its aspects. Other institutes, such as Rothamsted Experimental Station, are also on the trail. Chemists, too, are seeking better ways to extract nitrogen from the air than the Haber process, which is now half a century old.

Another far-out prospect is to extend our partial understanding of photosynthesis, the elaborate chemical sequence by which plants turn water and air into vegetable matter. Perhaps one day we shall have "artificial plants" which function even more efficiently as food-makers than natural ones. More plausibly, we may be able to improve the photosynthesis efficiency of the plants we have. (Some rather boring plants are highly efficient.)

Much nearer than these goals is the possibility of making edible protein from novel sources. Protein can be made from methanol or petroleum, and British Petroleum have a £30m.-plant making the animal-feed Toprina in this way. But oil is scarce, and such large installations are too costly for developing nations. Tate & Lyle, the sugar people, are working on a small-scale system for making protein from agricultural wastes, which would kill two birds with one stone.

There is also hope in the use of leaves and other vegetable matter, including grass which humans do not normally eat. Crude machinery for making leaf-protein was devised years ago by Dr Norman Pirie at Rothamsted Experimental Station, but the resources to develop it never materialised. Now that world-starvation is an active issue, development is being resumed.

Here, too, we can mention the texturing of soya-bean protein to produce meat-substitutes, such as Kesp, already in use as an extender in school and canteen meals.

And perhaps hydroponics will win its wings. This system of growing plants in tanks of water laced with nutrients has been around since the 1930s, and works rather well for high-value crops like tomatoes or carnations grown in greenhouses, where you want to cram a lot into the minimum space. Improved versions, developed by the Glasshouse Crops Research Institute, Bath University, and others are ironing out the drawbacks: a continuous flow of water suits the plants better than the static tanks of the past. Moreover, wind-power can be used to pump the water and to heat the greenhouse too, an idea now being tested in Dorset (see the colour picture on p. 83).

Agribusiness
Modern commercial farming has been dubbed, not altogether flatteringly, "agribusiness". It cannot be denied that it has brought lighter work, better pay and more security to farmworkers, without any proportionate rise in the cost of food.

However, there are limits to how far productivity can be pushed, and there may be long-term costs in land-impoverishment and social dislocation which could offset some of the gains. But Dr Kenneth Blaxter, F.R.S., director of the Rowett Research Institute, Aberdeen, says that despite the laws of diminishing returns "we are as yet nowhere approaching ultimate yields, or even reaching towards a point at which returns are negligible relative to effort . . ." As he sees it, the application of power to farming – not only power to drive machinery, but power in the form of manufactured fertilisers, power spent in making farm-machinery, and so on – constitutes a farming revolution. He sums up thus: "Whether there are still greater agricultural revolutions to come than the third revolution which I have predicted, and for which we must now plan, I cannot tell. No doubt most of us hope that Engels was right in stating that for science everything is possible. I believe that the third and subsequent revolution in farming can only come about through its continued application."

And now, to balance this optimism, let us see what Robin Clarke thinks.

The grape harvester is a
successfully exported cousin of
the blackcurrant harvester

Food and Farms

by Robin Clarke

We British farm pretty well. That is, the things that we do we do better than very nearly every other country in the world. We can out-produce almost any nation in intensive beef, milk, eggs, poultry and sometimes pigs. We can match the cereal yields of any of our arable competitors. But whether this holds us in good stead for the future is another question.

And it is a question which we ought to ask more frequently. I believe that our present mode of farming is living on borrowed time. It depends on cheap fuel and machinery – which are no longer with us. It depends on cheap imports of basic foods from other countries, particularly the Third World – who could put them to better use themselves. It depends on sustaining high consumption of animal products, such as beef, which lie a long way down the food chain – a remarkably wasteful way of using good agricultural land.* It depends, too, on keeping animals in confined situations. We claim to do this in the name of altruism – for the sake of producing more and cheaper food for a hungry world. In point of fact, we do it because it makes our farmers – of which I am one – richer. And, lastly, our whole system depends on dispensing with manpower on the farm and hence impoverishing our countryside – not to mention the majority of people thus condemned to live in towns and cities but who, when actually asked, mostly say they would rather live in the country.

Our farming may be efficient. But it is the type of efficiency which makes for an awkward world. Its awkwardness lies under the surface.

Farming is to do with two things: providing protein, and energy (in the form of carbohydrates). (True, our bodies need other things such as vitamins but these are either inevitable by-products or can be produced synthetically in a chemical plant.) How far does our farming succeed in this? We grow 50 % of the food energy we need and 62 % of the protein. (What we need is $\frac{1}{5}$lb of protein per head a day, or 70lb a year; and 2,400 kcal – "calories" – or 2.8 kilowatt-hours of food energy per head a day.)

*The food chain: Men eat cows which consume grass is a short food chain. We import 4 million tons of grain a year to feed the animals we eat. Editor's note.

Two Acres a Head

The half of this which we grow comes from our 46.6 million acres of agricultural land – a figure which includes about a third for rough grazing and hill and mountain land. As our population is now around 50 million, we manage to provide about half our food from just less than 1 agricultural acre per head. Each of us needs, in other words, 2 agricultural acres to provide all our food.

The proponents of the Western way of farming hold our example up to the world. But before doing so they ought to recall that the globe as a whole has only an average of 2.64 agricultural acres per head of population. Nearly all of that is vastly inferior land to England's green and pleasant pastures. Seen in that light, we are not doing too well – only marginally better than the world average.

The new style of farming in the West has certainly increased production. Since the turn of the century, in the U.K. we produce 1.4 times as much hay as we used to, 1.8 times as much barley, twice as many potatoes (in an ordinary year, of which we no longer seem to get so many!), about twice as much eggs, meat and milk and 2.2 times as much sugar beet.

But we do it by using more land. Back in 1900 there were no tractors, and all farm-work was done by horse (and the occasional steam-engine). Those horses solidly ate their way every year through the produce of one-third of our agricultural land. The tractor has put paid to such wastage and, though we have lost land to roads and cities, we still work about 28 % more land for human consumption now than at the turn of the century. When all that, plus imports of animal-feed, is taken into account, the actual increase in yield of food produced per acre on British farms since 1900 is not much more than 50 %.

Even that seems a lot. If all the world could do the same, many people would be a lot less hungry. But it is time to count the costs.

The Diminishing Returns

The farm used to be an ecological sort of a place. Roughly, what a farmer took from his land he aimed to put back in some other form. And he was always pretty self-sufficient – not only in the sense that he grew his own food (which of course he did; sugar and tea were the only imported food purchases on the nineteenth-century farm) but in the sense that

he bought very little from the outside world at all.
He could not afford to. His output was his profit,
save for his farm wages. And his fertiliser was farm
muck, spread on the land when the cattle were
turned out after the winter. Today muck is problem
number one for the British farmer. He no longer
knows what to do with it. It stinks and it pollutes. It
lies in huge festering piles all round the farm. And
the intensive farmer produces far more of it than he
could ever hope to spread on his fields.

This has come about partly through the invention of what in the trade is known as bag muck –
artificial fertiliser. Every farmer now buys it by the
ton. The use of artificial nitrogen doubled between
1900 and 1942. It had doubled again before 1950.
And again by 1972. We use 8 times as much as we
did in 1900. Even more lavishly applied has been
phosphate and potassium. Their use doubled
between 1900 and 1942. Doubled again by 1947.
Again by 1956. And again by 1965. Now we use
more than 30 times as much as at the beginning of
the century.

These huge inputs have increased the yield per

acre by only 50 %. And the rate of increase has
slowed to a fraction of what it once was. It looks like
the end of a road. The law of diminishing returns is
taking over.

The other big change to farming has been in
energy use. Where once we used men and horses
now we use tractors (and they unfortunately do not
reproduce but have to be made and purchased).
The food that each of us eats in a year can be grown
only by burning the equivalent of a ton of fuel oil –
in powering the tractors, the combines and the
other modern machinery of the farm. Perhaps it
does not sound a lot. Yet it is more than 3 times the
average world consumption of energy for all
purposes – heating, powering industry, cooking
and everything else. It is a monumental use of
energy. And what shall we do when the oil runs out?

In this respect most other farming communities
throughout the world and through history have
done better than we have. For example our
agriculture uses about 2.9 kcal of energy to produce
just 1 kcal of good energy. The ratio in the food
industry as a whole is 5 to 1. Broiler fowl need 10

A new British high-speed rotary
digger with a low-energy
requirement, designed for
primary cultivation

times as much energy as they contain.

Potatoes from Oil

Before mechanisation such a process would have been unthinkable. The whole point – or at least half the point – of growing food was to provide an energy surplus for the human body to consume. No primitive society ever put more energy into growing food than he got out of it. Some systems were indeed incredibly efficient when measured on this score. It has been estimated that Chinese peasants in the 1930s produced 40 times as much energy from their food as they expended in growing it. Value for money indeed.

Today, as one famous ecologist has put it, we grow potatoes made partly from oil. We no longer know how to do it any other way. We should learn. We may have to.

The law of diminishing returns works in other ways too. Farmers not only use bag muck but they feed their stock "out of the bag". Whereas they used to grow their own animal rations, now they have it delivered by the local corn merchant. Much of it comes from abroad. Much of that comes from the poor countries – in the form of fish-meal and soya beans, linseed and other high-protein foods. Most of that food could be fed direct to prevent starvation in the country where it was grown. Instead we buy it cheap and feed it to animals to provide the Third World with a cash crop.

Coffee used to be a cash crop of that kind. Now it is leaving the poor countries at a price of £4.50 per lb. We shall soon be forced to drink something else. And in the next decade imported protein will follow the price rises of foreign oil and coffee. And quite right too. But it will leave our intensive farmers in a nasty mess.

Finally, we should all think a bit about factory farming. I am no sentimentalist. I kill and eat my own animals. But take a walk through a battery hen farm, or an intensive piggery, and see how your stomach feels afterwards. Surely we have genius enough to improve on this?

What Should We Grow?

Western agriculture suffers from a very curious malaise. Farmers grow plenty of crops but they grow them for animals to eat, not people. Less than 9 % of our agricultural land is used to grow crops for human consumption; more than 91 % of it grows fodder crops.

This is not entirely stupid. In the hills and on rough ground, animals can make the best use of pasture. And however hard we try the human animal will never get accustomed to chewing grass, unless he grows another three stomachs! But grass is the principal crop on many of our most productive lowland farms; what is more, it is very often used to raise beef. Now the problem with beef is that it is a very inefficient way of producing protein. A bullock needs to consume 21.4 lbs of protein to put on 1 pound of its own protein.

Put it another way. You can grow 5 times as much protein per acre if you plant that acre with cereals than if you raise meat on it. That acre planted with peas or beans would produce 10 times as much protein. Just one acre of spinach, properly grown, produces 26 times as much protein as an acre supporting beef cattle.

But, farmers argue, beef is where the money is. They have a point, albeit not a very good one. The most damning argument I have ever heard of the British way of farming is that if the entire country were covered with council houses, from Land's End to John o'Groats, our agricultural productivity would actually rise! The reason is simple.

In 1951 two researchers set out to measure the productivity of the kitchen garden, or allotment. They found it produced a truly astonishing amount of food, worth a great deal of money. On average every acre of kitchen garden and allotment brought forth crops worth the equivalent of £42 an acre. In those days the best farm land produced crops worth £45 an acre, and the average value was only £36 an acre. Since on average only 14 % of the house plot is used to grow crops, it follows that the best thing to do with British farmland is to build houses on it!

Such frivolity hides a serious point. It is that the lower down the food chain we eat, the more productive – in terms of feeding people – will our agriculture be. Our problem is that animals provide 60 % of our protein and even 44 % of our food energy.

Our farm animals alone consume a resource which would feed 250 million people living off only cereals and vegetables. I do not suggest we should all become vegetarians. But I do suggest we would do much, much better to grow and eat more plant protein and less animal products – particularly beef.

The Social Tragedy

The real revolution in farming over the past half-century has been nothing to do with yields, improved crop varieties, more efficient milking-parlours. It has been a revolution of manpower, a flight of country people to towns and cities. Like most revolutions, its effects have been unexpected and in many ways undesirable: the countryside has been turned into a vast farming desert, swept relentlessly by gigantic machines, with too few people left to form the basis of a thriving country society. The cities have become over-populated, polluted and noisy. Most of us have the misfortune to live in the wrong place – which does not say much for our ability to engineer our social environment.

In 1885 the average farm worker looked after 26 acres of land – in my experience a reasonable allocation. Today he must tend on his own nearly 90 acres. No one can do it – that is, no one can do it well. By definition it has to be a rushed job, even with all the modern machinery in the world to help you. The way this has come about is shown by some revealing statistics. In 1901 there were 1.1 million farm horses and 1 million agricultural workers. By 1950 there were more tractors than men – we now run a fleet of 500,000 tractors while Bangladesh, a country the same size as the U.K., has to make do with 1,350.

And today about 400,000 people are employed directly on the farm – which is less than 2 % of our working population of 22 million. Each farm worker thus supplies about 66 people with the food energy they need, and somewhat more in terms of protein.

But that is not the end of the story. Altogether nearly 3 million people are employed in getting the food from the field to the consumer. Another million or so are employed overseas in providing us with the half of the food we do not grow. The total labour force is more nearly 4 million, and so each provides the food of 13 to 14 people. On average, we therefore have to spend 7 to 8 % of our working week in providing food.

Now this is undoubtedly better than other societies – but not that much better. In primitive society one person expected to provide the food for 5 to 7 people. For example, the !Kung Bushmen of the Kalahari desert – a very inhospitable environment – spend only 23 % of their working week in collecting food and hunting. They do it without any machinery, imports or other aids. Our capital- and energy-intensive processes have cut the time factor by three – but in the process have produced a chaotic social situation and a hideous urban problem. No one can be blamed for asking whether it has all been worth it.

What Shall We Do?

Even so it is the wrong question. The right one is what shall we do about it? And here British Genius can surely provide some new answers for a new future. We need smaller farms, more highly staffed, producing more food on fewer acres – and that means more protein per acre, not more profit per acre.

We need to bring the food-processing industries back from the urban environment and onto the farm and village where they can be run by farming co-operatives to swell the country population and take the pressures off the city. We need methods – such as the anaerobic digestion of farm manure to produce methane – to make farming once again energy independent. And finally we need fiendishly sophisticated technology which will bring in our farm crops with cheap, easily maintainable and simple machinery. That machinery will need more men to run it than do the current models – but the work involved must not include the back-breaking drudgery of former agricultural periods. Energy consumption must be small.

We should not seek, as Chairman Mao once remarked, to eliminate the difference between town and country. But we should redress the balance – particularly where the balance is so unfavourable to both sides. To do all this will in no sense mean turning our backs on progress. We need, merely, to progress in a different way. In the industrial revolution we British once boasted we were the first country in the world to get off the land. Let us hope we can be the first one back again.

This rotary unit for vegetable preparation reduces labour on pre-packing by about 30%

Challenge No. Three

The More The Merrier?

The supply of food and the availability of energy are our basic challenges. It seems that there will be enough of both – in the next quarter century at least – but in meeting these demands we may radically influence our lifestyle and the pattern of the way we live. When we turn to technological development in general, we uncover the same underlying problem. Do we just want more and more of the same kind of progress; primarily improvements in communications and transport? As the British Genius Exhibition shows, our inventors are displaying enormous ingenuity in developing new forms of familiar things: we shall move people and ideas even more freely. But they are doing little to counter the problems – such as noise, congestion, rootlessness and social disruption – thrown up by those inventions. They are doing even less towards building the alternative society which some people would like to see, a society more decentralised and based on self-fulfilment rather than mass consumption. By the same token, they are doing little for the Third World, except to encourage it to follow in Western footsteps.

But we can only chronicle British genius as it exists; here Dr Tom Margerison, the Exhibition Editor, surveys some of the outstanding devices seen in the Exhibition and sketches in the background of industrial effort which lies behind many of them.

Womb of Time

by Tom Margerison

In the field of general technology – transport, communications and suchlike – British inventive genius is seen to be alive and well and visible in Battersea. The biggest single influence on technology at the present time is the application of solid-state electronic techniques to every kind of industrial process and activity, and in particular the use of the information processing developed by the computer industry. The new methods are now being used not only in the operation of plant, but also to control it. As these methods (to which we shall be returning later) are introduced more and more widely, they generate a new demand for telecommunications. Britain has developed a gas-

Britain plays her part in space,
building satellites, here Ariel IV

The Post Office Tower seen
through the trees of a
neighbouring square

grid conveying natural gas from the North Sea to users all over the country. In order to control the grid, continuous measurement of flow-rates in different parts of the system have to be made, and are sent over telephone lines to a computer centre where the whole network is analysed and commands sent out, again over telephone lines, to operate pumps and valves.

These new applications are going to call for much greater telecommunications capacity in the future. But existing cables are limited in the amount of information they can carry, and the microwave link system, based on the Post Office Tower in London which interconnects with other towers in different parts of the country, is also reaching the limit of its capacity. Post Office engineers and the telecommunications industry are looking for other ways of conveying economically enormous amounts of information.

There are two contenders, both of which are in development. The first, now at its prototype stage at the Post Office Research Station at Martlesham Heath in East Anglia, is a waveguide. It can be regarded quite simply as a pipe through which radio waves are squirted. In fact one of its ardent proponents, Professor Harold Barlow of University College London, has demonstrated a waveguide carrying telephone conversations, electrical power and gas simultaneously! The problem with waveguides used to be that, although they could carry many thousands of telephone conversations, they needed to run straight if losses were to be kept low. That problem has been overcome by a new design of pipe, and about ten miles of prototype waveguide have been laid between Martlesham and Wickham Market, in Suffolk.

Voices in a Glass Thread

The second approach, which is currently the fashionable favourite, is even further removed from the familiar telephone cable. Two engineers working at Standard Telecommunications Laboratories at Harlow – Charles Kao, a native of Hong Kong, and George Hockham – suggested that instead of using radio waves in a waveguide, light could be used as the communications medium, conveyed in a glass fibre which might be thinner than a human hair. The suggestion did not meet with much enthusiasm. Glass available in the 1960s was insufficiently transparent; there was no

suitable light source available which was both reliable, relatively cheap and uncomplicated, and could be switched on and off rapidly enough to send the sequence of flashes which made up the encoded information. But Kao and Hockham persevered. The glass was improved, until today a tiny pulse of light sent into one end of a six-mile-long glass fibre can be detected without difficulty at the other end. The glass is so pure that it would be possible to see through a six-mile-thick window made of it.

Within a few years of that seemingly impractical proposal, new infra-red solid-state lasers were developed which would produce a rapid succession of pulses of invisible light without fuss or bother. In fact, they are little more difficult to handle than an ordinary transistor. With these two components now available, the opportunities for glass-fibre communication seem very good. The techniques for making cobweb-thin fibre into robust cables clad in plastic are being worked out; new ways are being found for splicing glass fibres without losing the light flashes they are carrying. At present each fibre can carry 10,000 telephone messages, or several complete television signals, and its capacity is likely to be increased still further as development proceeds.

The impact of optical fibres will not be felt very quickly because the investment in the conventional telephone system is enormous: the changeover from copper wires to glass fibres at subscriber level would be impossibly costly. But it should happen in

Experimental manufacture of glass fibre for fibre optics: a glass rod is heated in a gas ring and emerges as the finest thread

Optical cable (in the hand) is compared with traditional copper cable (top left) and modern coaxial cable: the optical cable can carry more messages than both the other cable systems

the near future on trunk lines, particularly as the special glass will probably soon be cheaper than copper. Glass, moreover, is made from raw materials which are much more plentiful than metals.

When the glass-fibre revolution reaches the home, it will immediately make possible all kinds of services which cannot be considered now because a telephone connection carries so little information. It will mean, for example, that the video-telephone could become a reality, allowing a subscriber to call up on a screen his own choice of newspaper or book. It would give him access to data-banks containing any kind of information – from telephone directories to shopping price-guides. It might make it possible for him to conduct his business without actually leaving his own home. In the Exhibition there is a demonstration of how a television picture is sent over one of these glass fibres. What was an entirely British idea is now being developed all over the world.

The Problem of Moving People

While in theory new ideas in telecommunications could reduce the amount of travelling that is done, they are hardly likely to affect the demand for transport either in towns or between cities. Urban transport is rapidly becoming a critical problem as outdated street layouts are progressively less able to cope with the onslaught of traffic. Some modest palliatives are becoming available through the development of computer-controlled traffic lights. Recently, a development of the Plessey Company, which enables vehicles to identify themselves on being interrogated by induction coils laid under the road surface, opens up new possibilities for controlling and segregating traffic. Eventually a system could identify any vehicle on the roadway and charge the owner with the cost of using the road. In the shorter term, the system will be used for identifying buses in order to monitor the service and to give them (and other special vehicles such as ambulances) priority at intersections.

Passengers are transported at
Dorval Airport, Montreal, on
Dunlop Starglide belts

In the long run, the centres of towns need to be redesigned to favour public transport. An interesting approach to this problem has been used in Runcorn New Town, where a network of reserved bus-roads has been built to carry people to and from the shops and factories. Such developments are only possible when a town is being newly built. An alternative is to build a separate public transport system on its own reserved track, like the London Underground system. The problem is cost. The systems which now seem most favoured, both in Britain and elsewhere, are based on small trams powered by electricity. One new approach is the "magnetic attraction" tram designed by Professor B. Jayawant of Sussex University. The tram has no wheels, but is suspended from its tracks by means of electromagnets. The secret is the control, by sophisticated yet inexpensive electronics, of the current through the magnets. Only one kilowatt of electric power is consumed in the magnets for each ton supported. As the trams have no wheels they are silent, and track maintenance is also reduced.

The "maglev" train has been much discussed as a higher-speed successor to the inter-city train, and several ingenious systems have been developed by British engineers. At present most of the interest is focused on very-high-speed trains using super-conducting magnets to levitate them above the track. These super-conducting magnets produce extremely powerful fields and use very little electrical power, but they do need to be cooled to temperatures not far short of absolute zero. The saving in electricity is thus offset by the cost of the refrigerating system needed. Nevertheless, three groups, led by Professor Eric Laithwaite of Imperial College London, by Dr Rene Rhodes of Warwick University, and a group at the Rutherford Laboratory of the Science Research Council, are working on the design of vehicles of this kind.

Vertical Take-off – if you want it

The difficulty with these very-high-speed land-based vehicles is that they have to compete with already well-established airlines. It seems doubtful

The Rolls-Royce Flying
Bedstead rises

whether the attractions of the very-high-speed super-conducting train will prove sufficient to make it an acceptable alternative to the airliner. Even new designs of aircraft have difficulty in competing with the old if different airfield requirements are called for. For example, it would now be technically possible to develop a vertical take-off airliner which could fly from city centre to city centre, would create less noise than conventional jets, and would require only relatively small landing pads. Hawker Siddeley has carried out detailed assessments and design studies for an airliner powered by 2 conventional fan-jets for forward flight and 16 small lift-engines. The lift-engines, designed by Rolls-Royce, would be built largely of plastics. Britain has always been in the forefront of this work, ever since the first experiments with the Rolls-Royce Flying Bedstead in the 1950s. But the vertical take-off airliner seems unlikely to be developed simply because the investment in long runways for conventional aircraft has already been made.

The Microprocessing Dilemma

The advances in electronics in the past thirty years have been called the greatest technological advance the world has ever seen, but the great surge in the development of equipment based on electric controls, computer technology, and the use of television-display techniques has still not reached its peak. The development of microprocessors, in effect complete mini-computers printed on to a single chip of silicon, is making the application of the new techniques even simpler, for instead of designing special-purpose electronics the engineer can now fit a standard microprocessor. The microprocessor is then "programmed" with a series of instructions to make it perform in the way required. The advantage of this method is that the programme of instructions can be developed and tested independently of the equipment itself. When some change is necessary to upgrade performance the programme can be changed without the equipment needing to be touched at all, and the new instructions fed in a matter of minutes into every machine. Updating equipment in the field thus becomes very simple and inexpensive.

To market these tiny yet complex devices, each containing tens of thousands of active elements, calls for a massive investment. But the more

elaborate the device, the more specialised and the more limited the market. This is the manufacturer's dilemma and explains why the majority of microprocessor "chips" are manufactured in the United States. But at least one British firm, Ferranti, has in the last few months put a particularly powerful one on the market which should find many industrial applications both in Britain and overseas.

How to See in the Dark – and the Wet

Television technology is also being used in a wide range of industrial equipment. Miniature camera-tubes, only half an inch in diameter, are used for looking down pipes and into chemical plant to check the extent of wear and corrosion. New types of television camera-tube have been developed which are sensitive to the far infra-red, and to sound. The infra-red camera, which is based on work carried out at the Royal Signals and Radar Establishment in Malvern, detects the heat radiation from objects at around room temperature. It produces brilliantly clear television pictures, even sharper than broadcast television, but taken entirely by the heat from the objects, and in absolute darkness. The lenses for this camera are made from polished germanium. Pictures of the countryside at night, taken with the infra-red camera, show foxes stalking rabbits. The plumes of steam at a power-station appear brilliantly white, while the cold-water pipes leading to the cooling-towers look black.

So far, the major applications are in defence, but the cameras obviously have wide uses in industry, for security, and in medicine. They may even be useful for archaeological and agricultural surveys, as objects hidden beneath the earth affect the soil temperature, as do different types of crop. Clothes are transparent to infra-red rays, so the use of the infra-red television camera for human photography can be somewhat revealing.

The sound-sensitive television camera is a still more extraordinary device, developed by the Central Research Laboratories of E.M.I., where the development work on the original B.B.C. television system was carried out. The whole camera is filled with water, and it works only under water. The camera-tube is similar to a normal television camera-tube, but it has a target which is made of a very thin layer of quartz mounted on a plastic plate.

If sound-waves are focused on to the quartz, a pattern of electric charges forms on it, and these are read off by an electron beam. The sound-waves are focused on to the camera tube by a special lens made of polystyrene.

This device enables objects to be seen in the very muddy waters of the North Sea, where normal visibility is zero. Special ultra-sound "floodlights" are used to "illuminate" the objects to be examined with sound. The sound-waves scattered from the objects are focused by the lens on to the camera-tube, and these are then displayed on an ordinary television screen. The quality of the pictures is not as good as a normal broadcast picture, but the links of a chain or the fingers of a hand can be seen clearly.

Swings, Roundabouts – and Buses

Not all the developments for the future will be electronic. Some of the most elegant are mechanical, although the development period for a major mechanical project tends to be many times longer than for an electronic one. Donald Firth, of the National Engineering Laboratory at East Kilbride, has spent twenty-three years working on the design of the ball hydrostatic motor, which is now being used to power all kinds of vehicles from small cars to heavy contractors' machines. But its real potential

has yet to be exploited. One interesting application is to provide a regenerative braking system for city buses. When a bus accelerates from a stop, it uses fuel energy and converts it into the energy of motion. As it brakes to stop again, this energy of motion is dissipated as heat in the brakes. Between 60 and 70 % of the total energy is lost in this way on a typical city bus-route. Why not save the energy lost in stopping, and use it to get going again? Donald Firth is now fitting his ball hydrostatic motors to the wheels of a bus. These motors run on high pressure oil fed from a pump coupled to the engine. The hydraulic system acts as an automatic transmission. But most important, by changing the timing of the valves the hydrostatic motors can be used as pumps, pumping the oil into an accumulator where air is compressed. The operation of pumping brakes the bus. The result is that the energy of motion of the bus is converted into the energy of compressed air in the accumulator. The compressed air acting on the oil then helps to accelerate the bus as it moves off again. Even without the engine running, the bus would be able to recover enough energy to accelerate to about 15 miles an hour. It seems likely that between a third and a half of the fuel used in buses could be saved in this way.

This photograph of a lobster
boat taken at night, shows off the
infra-red camera

The Simplicity of Genius: Nets from Nozzles

A similarly elegant mechanical solution to a problem has given a small British firm a unique product that sells all over the world. The idea is of such dazzling simplicity it seems extraordinary that nobody thought of it before. Brian Mercer, who has a small family textile firm, invented and developed a machine which extrudes not rods but plastic net. Nets are normally made by knotting yarn or weaving wire. Mercer's process consists of extruding all the threads in the net together through concentric counter-rotating heads. The parallel threads come from the outer head moving (say) clockwise, and from the inner head moving in the opposite direction. The holes through which the threads are extruded are machined as slots on the inside of the outer ring-shaped head, and on the outside of the inner cylindrical-shaped head. The two heads fit exactly into one another. As they revolve in opposite directions the slots on the inner and outer heads momentarily coincide, extruding a double thickness thread which is the cross-over point of the net. The result is a continuous tube of net, whose size, mesh and thread thickness depend on the size and design of the extruding heads, and the speed with which they are rotated.

Netlon, Mercer's firm, not only markets a wide range of nets for all kinds of purposes from supermarket food packaging to anti-dazzle screens on motorways and soil reinforcement, but he has been able to license Netlon to plastics extruders all over the world. The firm recently won a Queen's Award to Industry for technical innovation, and

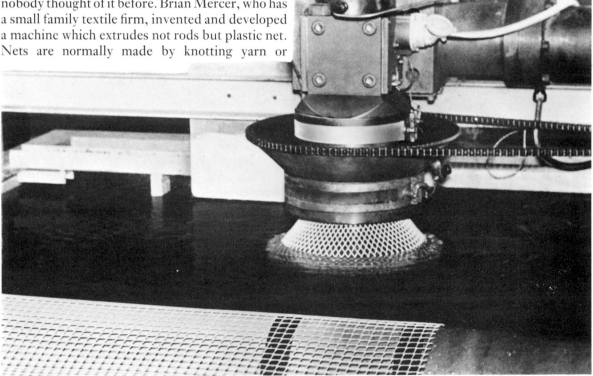

The basic ball hydrostatic motor
designed by Donald Firth

Newly-made net passes from the
Netlon extrusion machine into a
cooling tank

Mercer the Royal Society's Mullard Award. In this case, success came not simply from a brilliant idea, but one which it was within the capability of a small firm to develop, which met a clearly defined and readily appreciated market need, and which could be applied speedily and at low capital cost by thermoplastic extruders anywhere in the world.

However, things do not always go as smoothly as this, and it is worth looking at two stories which exemplify some of the difficulties firms face in marketing new ideas.

The Float Glass Saga
and the Rolls-Royce Crash

Sir Alastair Pilkington noticed that during washing-up fat congealed on the surface of the water to form a thin sheet with polished surfaces. He saw that if molten glass could be poured on to a liquid surface and allowed to solidify it might be possible to make sheet glass with a good optical surface without the need for the expensive mechanical grinding and polishing processes needed for plate glass. His first experiment, in 1953, looked promising. But the development cost £7m. before it had reached the stage of commercial application. By the time the first successful sheet floated out of the pilot plant on a pool of liquid tin, £4m. had already been disbursed unsuccessfully. The Pilkington story is one of success. The risk paid off, and the company, which already had the means of marketing plate glass, was able to launch the superior and cheaper float glass. The process has now been sold throughout the world.

Less happy is the story of the development of the RB-211 jet aircraft engine by Rolls-Royce. The idea was to build a highly economical engine of very large size suitable for the new wide-bodied aircraft such as the Boeing "jumbo" jet. The design involved several new concepts. First, it included an enormous fan, able to handle six times the quantity of air required for the gas turbine. The idea of using an over-large fan as the first stage of the compressor

in the jet engine, and then bypassing the extra air round the gas turbine to mix with the hot jet was a brilliant development which had been introduced by Rolls-Royce in the mid-1950s in the Conway jet engines which powered the V-bombers. The cold air from the fan mixing with the hot air from the engine cooled and slowed down the jet, giving the aircraft greater thrust for a given amount of fuel burned. The engine efficiency could be made still higher by increasing the size of the fan. To do so meant using three separate turbines, working at different speeds, one to drive the fan, one for an intermediate compressor and one for a high pressure compressor. In this way not only could more air be bypassed round the engine to join the jet, but each compressor could be run at its optimal speed, and the maximum compression obtained, which again contributed to high efficiency.

The problem was how to build such a large fan with the tips of its blades moving at close to the speed of sound. The centrifugal forces on the blades under these conditions is enormous, and an extremely strong, light material is therefore needed.

The Rolls-Royce engineers looked with interest at what appeared to be the ideal material, which had been developed by two engineers – Leslie Phillips and Bill Watt of the Royal Aircraft Establishment, Farnborough – in 1963. They had discovered that if they carefully charred synthetic fibres, such as Courtaulds' Courtelle, in which long molecules are neatly arranged in parallel lines along the fibre length, the black silky strands of carbon had remarkable properties. During the charring process the carbon atoms retained their alignment and the result was fibres far stronger than the strongest steel. By incorporating these fibres as reinforcing materials in a plastic matrix, Phillips and Watt produced a material four times stronger, weight for weight, than high-tensile steel, and much stiffer than the glass-fibre reinforced plastics.

The material seemed ideal for the blades of the

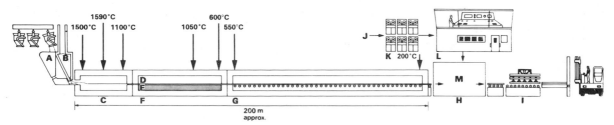

Pilkington's float glass process:

A Raw material mix	F Float bath	J Orders
B Cullet	G Annealing lehr	K Computers
C Oil-fired melting furnace	H Automatic warehouse	L Control point
D Controlled atmosphere	(not shown)	M Computers govern the cutting
E Molten tin	I Automatic stacking	processes: the automatic

warehouse stands by itself as a major advance in flat glass technology

giant RB-211 fan. Calculations showed that it should be possible to save 300 lbs weight per fan compared with one built with titanium blades, which cost more and are harder to machine. When tested, the new engines built with the carbon-fibre fans met the engineers' expectations. It was only when development flying started that a literally fatal flaw appeared: something quite unforeseeable. In any jet aircraft birds are occasionally sucked into the engines – and the carbon blades were shattered when this occurred. Production came to a halt. This costly setback contributed to Rolls-Royce's financial crisis, which obliged the Government to nationalise the company. But since this crisis the RB-211, now fitted with composite titanium and carbon-fibre blades, has proved a tremendous success. It powers the Lockheed Tristar airliner and is being fitted to other large aircraft. Yet its partial failure was enough to bring a large firm to its knees.

Nevertheless, the future for carbon-fibre reinforced plastic (C.R.P.) looks good wherever a combination of high strength and low weight is essential. Already it is being used for the cowls of aircraft engines: indeed the prophets of the Royal Aircraft Establishment claim that the aircraft of the future will be not silvery in appearance but black! There are many more ordinary engineering applications for C.R.P. It could replace steel in the manufacture of gas cylinders and pressure vessels. It would have advantages in the unsprung parts of vehicle suspensions, in making machine-tools, and in rapidly rotating machines. On a smaller scale, it has already proved itself to be the best material available for the shafts of golf-clubs.

These stories illustrate an important point. If, in the British Genius Exhibition or elsewhere, you see devices which strike you as brilliantly inventive, do not at the same time underrate the skill, courage and persistance of those who succeeded in bringing them to marketable form or are still trying to do so. As well as the ideas which are noticed, there are others perhaps just as good which are not taken up. This question of exploitation will be considered in more detail in the last chapter, for it is crucial to our future.

Challenge No. Four
Your Good Health!

Dr Robert Heller, then the director of America's National Cancer Institute, once asked me how it was that British researchers produced half the good ideas on cancer research when the money spent was only one-fortieth of what was spent on such research in the U.S.

The fact is that, ever since scientific medicine emerged from the mists of superstition, the British – and especially the Scots – have contributed much more than their share to medical advance. And that state of affairs shows no sign of changing. In recent years, for example, British workers have discovered an entirely new agent which can be used against viral diseases: interferon, discovered by Alick Isaacs and H. A. Lindemann in 1957. It has proved very difficult to make in quantity but recent trials at the Cold Research Unit suggest that it may be the long-desired cold cure. Furthermore, it seems to be not so much an anti-viral agent as a control substance for the cellular machinery and may prove of quite fundamental importance.

It was Britons too who evolved the newest antibiotics, the cephalosporins, described as the "antibiotics of choice for the 1980s".

We are outstandingly good at devising medical equipment. From the use of radioactive gas to study lung functions to the E.M.I. scanner, seen in the Exhibition, and the Diasonograph, which uses sound waves, British medical equipment is used all over the world.

British doctors are among the leaders in the field of surgery. The "Charnley" hip replacement for arthritic hip-joints has been copied all over the world. British shoulder protheses are also winners.

In all, Britain spends at least £100m. on medical research – not counting what is spent by the pharmaceutical companies.

I asked John Newell, the medical correspondent of B.B.C. External Services and contributor of medical notes to the *Annual Register*, to summarise some of the exciting medical advances which are coming into use.

British Genius in Medicine

by John Newell

In 1976, the *American Journal of Roentgenology* devoted its entire bicentenary issue to a British invention, justifying this decision by saying that it was the most important advance in the field of diagnosis since the invention of X-rays. The invention which evoked this panegyric was the brain and body-scanner developed by E.M.I.

Anyone who has studied botany at school will remember slicing cross-sections of plants and mounting them on slides for examination, so that the various structures and their relationships were clearly displayed. It would be convenient for doctors if they, likewise, could slice their patient across from end to end and examine the more interesting slices. This is virtually what the technique known as Computerised Axial Tomography (C.A.T.) achieves. Beams of X-rays are shot through a "slice" of the body from 18 points, then a computer analyses the data to build up a picture of what the cross-section would look like (as you see in the colour photograph on p. 93).

Moreover this is done so rapidly that, if the slice does not reveal enough or has been taken in the wrong place, further slices can at once be taken. The applications are numerous. Lung disorders, such as tumours and infections, are clearly revealed; as are blocked air-passages and blood-vessel damage. Inside the heart, abnormalities of the valves and chambers can be seen. In the abdomen, liver cysts, gall-stones, bowel tumours, stomach ulcers and so on are clearly displayed. Even in the brain, blood-clots and tumours can be visualised. Recently, the method has shown physical abnormalities in the brains of severely schizophrenic patients. Moreover, the effects of treatment on any of these conditions can readily be monitored.

The first C.A.T. scanner (designed by Geoffrey Hounsfield of E.M.I., Electric and Musical Industries) was introduced in 1972, after five years of research. It was intended to scan the brain, being too small to admit the whole body. In 1975, the whole-body-scanner was launched and recently an updated version of the brain-scanner, providing more detail, was introduced. More than 400 of the brain-scanners and 250 body-scanners have already been sold, earning £125m. in foreign exchange, and half a dozen American firms are desperately trying to climb on the bandwagon by

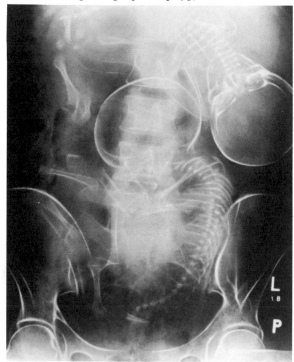

An X-ray shows twins in their mother's womb

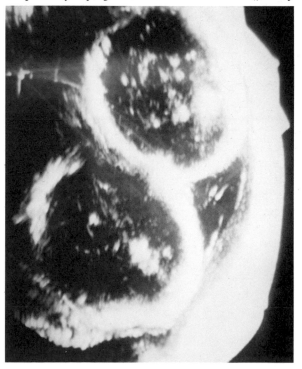

The heads of the same twins are revealed by ultrasonics, using the Diasonograph

producing something similar.

In the body-scanner, the patient is surrounded by a scanning gantry, carrying the X-ray sources, while sensitive X-ray detectors are positioned on the opposite side. A single scan, which involves the source moving through 180° and making 18 exposures, takes only 20 seconds, and covers an 18mm-thick slice. The computer stores about 80,000 readings which are assembled into a picture, seen on a television-type screen. The tape, or disc, carrying this data can be stored for future reference. After each slice, motors advance by 13mm the table on which the patient lies for a new survey.

The scanner can be linked to a machine which inspects X-ray pictures of tumours and designs the most appropriate radiation therapy for them; the first set of this kind has recently been installed at a south-coast hospital. This avoids the need for an exploratory operation and, in addition, the machine can calculate how far intervening body tissues will attenuate the therapeutic X-rays and make due allowance for this in calculating the dose.

The C.A.T. scanner has one slight drawback: it involves the use of X-rays and all X-rays involve some danger. The dose is not large, and is acceptable for a person who, if not treated, might otherwise die. Nevertheless, a system which does not use X-rays is to be preferred when possible. Such a system is the Diasonograph which relies not on X-rays but on ultra-sound. There can be no doubt that in medical ultrasonics the British are pioneers and are far ahead of the rest of the world.

While most diagnosis depends on a single examination, there are disorders in which it is really necessary to study how the patient's condition changes throughout the day. It is not always very helpful to measure a man's heart-rate while he is in the surgery. You want to know how much it shoots up when he is overtaking another car, arguing with his boss, or having a nightmare. Heinz Wolff of the Clinical Research Centre has designed a tiny tape-recorder which runs for 25 hours and can be carried around all day, continuously recording heart-rate, brain-waves or other wanted information. As he points out, you cannot tell much about the outcome of a football match from a still photograph, but you can tell a lot from a film of the entire game. Epilepsy is another disease where such a device is useful, revealing how often, and when, abnormal brain-activity shows signs of developing.

Computer analyses of the recording have revealed variations in the performance of the heart, the very existence of which had not previously been suspected. Ambulatory monitoring, as it is called, is growing in importance and before long no major hospital will be without a department to conduct it.

The computer has also transformed another aspect of patient care: the examination of specimens of blood, urine and other body fluids. Formerly done in a pathological laboratory by technicians, doctors often had to wait for days for a report. Now, in modern hospitals, the whole process is computer-controlled, the tests being performed automatically. Only the significant results are offered to the doctor and they come in a form easy to understand, and within 24 hours. In a hospital so equipped, half a million tests can be performed each year.

Biological Methods

But advances in diagnostic methods are not always due to sophisticated hardware. British medical workers have been evolving some subtle new methods of a more biological character. Another technique, developed at the Hammersmith Hospital, uses human white blood cells which have been made harmlessly radio-active as "hounds in the blood" to lead doctors to the site of hidden infections. The technique has already been used to locate deep, hidden abscesses, so that they can be drained. Since different types of white cells are adept at locating different forms of illness, the doctors plan to use various classes of "bloodhound" white cells to locate tumours and blood-clots as well as deep-seated infections. (And I must at least mention the rival techniques of radio-immuno assay and enzyme-linked immunosorbent assay – both of which have been very largely developed in Britain – which allow the presence of infection, or of drugs or other compounds, to be detected infallibly from tiny samples of a patient's blood.)

Delivering the Goods

Directing drugs and radiation to tumours is the next step after locating them. Research workers at the Tumour Immunology Unit of University College London have identified substances, proteins, found only in cancer cells, leukaemic cells, and, although only in laboratory experiments so far, they have created drugs which react specifically with, and attach themselves to, such compounds as a key fits a lock. They hope to develop drugs which will be targeted to cancer cells in this way, so that a much higher proportion of the drug affects the cancer cell than is the case today, thereby giving a much higher chance of a cure. The same principle is being developed to attack other cancers.

Drugs can be made more effective by protecting them from breakdown on the way to the tumour. Tiny droplets of oil composed of several layers, like miniature onions, with drugs sandwiched in between, have been invented and are being developed in several British laboratories, as a means of delivering drugs to their targets more effectively and specifically. These micro-pills, known as liposomes, can be coated with compounds which lead them to react specifically with tumour cells, and so to deliver the drug inside only to such cells.

The majority of cases of human cancer are now thought to be due to environmental factors, mostly chemicals in the environment we eat, drink, touch or breathe. Even those human cancers which are now thought to be caused by a virus probably require some environmental factor, to activate the virus. So we might achieve vast reductions in the incidence of cancer by devising sensitive tests to screen out carcinogens. One of the few widely applicable, practicable and sensitive of such tests has recently been developed by Dr Jim Bridges of Surrey University. It involves mixing a small amount of the chemical under test with animal liver cells in the laboratory, and then testing to see if the liver cells behave differently, in the chemical reactions they themselves perform, as a result. Tests like this could reduce the incidence of cancer by perhaps as much as 50 % if the carcinogens they identified were to be excluded rigorously from the environment.

Yet another hopeful approach to cancer lies in finding better ways to stimulate the latent immune response of the patient's own body against the cancer. After early bright hopes, progress in this direction has been rather disappointing. But, by identifying the exact nature of the immune (defensive) response to some tumours, and developing new drugs specifically to stimulate this part only of the body's natural defences, research workers at the Clinical Research Centre, London, have developed a hopeful accessory treatment for

some cancers. It uses a starchy compound, called C.O.A.M. for short, which stimulates one class of white cells in the blood, called macrophages, to join the battle. C.O.A.M. is no cancer cure, or even a treatment on its own. But the solidly-founded hope is, that by allowing stronger and more effective stimulation of the macrophages than has hitherto been possible, C.O.A.M. will help the patient's body to destroy cancer cells remaining after major treatment by drugs, surgery or radiation, and so reduce the chance of the cancer growing again.

The No. 1 Killer

There is no doubt that the Western – and East European – epidemic of coronary heart disease is due in the main to a way of life, among men at least, far removed from that which the human animal evolved to fit. Mild exercise, the avoidance of stress, a sensible diet and giving up smoking can do more to reduce the chance of a coronary than any amount of drugs. But drugs have their place.

Within the last few months a team led by Professor John Vane of the Wellcome Foundation's laboratories, at Beckenham in Kent, have reported their discovery of a naturally occurring hormone which inhibits the clotting of blood. The newly discovered hormone is one of the group of body chemicals called prostaglandins. Not only may this give us an anti-clotting drug to avoid clots forming after operations, but because prostaglandin X, as the compound is being called, appears to be the natural means by which arteries prevent the formation of thickening deposits upon their walls it may also prove possible to develop the hormone into a drug to protect against coronary thrombosis.

Many people, through no fault of their own or their life-style, are especially at risk of coronary heart disease, because they suffer from an inherited condition called familial hypercholesterolaemia. Between 50,000 and 100,000 people in Great Britain alone suffer to some extent from this condition, which causes high levels of fatty cholesterol in the blood. At the Hammersmith Hospital Dr Gilbert Thompson is treating such patients by a unique "blood-swap" technique he has devised; their blood is centrifuged, the blood cells being saved and fed back into the blood circulation, together with fresh blood plasma of the right group containing no cholesterol.

The patient's own plasma goes to a medical centre, where the cholesterol is removed and the blood stored to be recycled into another patient. Regular blood swaps every few weeks are reducing such patients' risks of coronaries to normal levels.

The blood-swap technique, developed wholly in Britain – though now being taken up elsewhere – may ultimately become as widely used as dialysis, the use of the artificial kidney. It is already being used to rid the blood of other dangerous compounds, which cause kidney disease and the muscle-weakening disease myasthenia gravis.

The value of proper diet in preventing heart attacks is accepted. But a scientific definition of proper diet is required. Dr David Jenkins and his colleagues, who have moved into the kitchens of the Central Middlesex Hospital in West London, have learned how to incorporate the vital constituents of fibre, plant-cell walls, into ordinary foodstuffs. They have shown that their specially made marmalade and bread can cut blood cholesterol levels.

Another food research centre, the Dunn Institute at Cambridge, is working with Unilever to develop concentrated extracts of the same valuable compounds, pectin and guar, found in plant fibres. The outcome should be foodstuffs which put the missing vegetable-diet-type chemicals into Western diet without enforcing a changeover to unacceptable or unavailable foodstuffs.

Relieving Pain

Intractable pain, due to arthritis, nerve disease, cancer or other causes, affects many thousands and not just among the elderly. Britain's specialist Pain Relief Unit, at the Abingdon Hospital near Oxford, has developed a unique pain-killing procedure, a "cold probe" which is used to destroy a section of pain-carrying nerve by momentarily freezing it. The technique has big advantages over the continued use of addictive drugs; it completely eliminates pain for up to half a year and has proved so effective for treating chronic, intractable pain that it is now being used to abolish the brief period of serious pain after some surgical operations.

The great problem in pain relief by drugs is that of addiction. One of the most massive international research operations of all time, which originated from a British discovery, and in which British scientists remain among the front runners, is now moving towards the development of a non-addictive pain-killer as powerful as, perhaps more powerful than, morphine. The new drug will be developed from a naturally occurring brain-chemical christened enkephalin, discovered, identified, isolated, analysed and purified in little over a year by a team of scientists in Aberdeen, Hull and London Universities, led by Dr John Hughes of Aberdeen. Besides affecting the sensation of pain, enkephalin seems to control mood, and to induce states of happiness. The enkephalin story grows more complex every day and the part played by this substance or group of substances in the brain, and their potential for medicine appears greater all the time.

Blue Babies

Heart-pacemakers, giving a regular electrical stimulus to regularise the beat of a flagging or irregular heart, are well established. Their snag is that, unlike the natural pacemaker, they do not adjust heartbeat to the varying needs of the body. Sussex University researchers have now developed a pacemaker which takes its own cue from the small heart chambers, the atria, which fill the big pumps, the ventricles. In this way the pacemaker follows the need of the body.

A Scottish chemical engineer, Dr Norman Macleod of Edinburgh University, has designed a

A version of the Edinburgh artificial heart valve, made from plastic and stainless steel

completely new form of artificial heart valve, made from carbon, which reduces the risk of clotting, gives better performance than conventional valves, and is also cheaper to make.

A new operation to correct a major birth abnormality of the heart has been pioneered at the National Heart Hospital. This is the kind of "blue baby" defect in which the main arteries leaving the heart are misplaced and swapped round, so that blood already exhausted of its oxygen is pumped round the body yet again, while blood which has just returned from the lungs full of oxygen is sent there once again, instead of travelling to the oxygen-starved body.

In the past this has been only partially corrected by surgery, by swapping round the misplaced veins. At the National Heart Hospital they have now perfected a much more difficult operation, in which the arteries leaving the heart are cut at the roots, so to speak, swapped around and sewn back in the positions in which they should be. The operation also involves cutting and exchanging the vital coronary arteries supplying the heart itself, which are little bigger than bootlaces in infants on whom such operations are performed. The procedure leaves the heart exactly as it would have been in a normal child.

Histamine and Ulcers

Ulcers, like coronaries, are a stress condition on the increase in Britain. They are common, painful and hard to get rid of. Treatment has been generally unsatisfactory, acid-neutralising compounds give only brief relief and surgery is dangerous. But, over the past few years, Professor J. W. Black, working with drug companies and at University College London, has developed a completely new treatment for gastric or duodenal ulcers which has been completely effective in about three-quarters of the cases for which it has been used. This is no less than a revolution in ulcer treatment – though, of course, it is essential that the patient thereafter avoids a return to whatever stressful conditions or habits led to the appearance of the ulcer in the first place.

The first drugs based upon Black's research have now become commercially available in Britain, for the first time anywhere in the world. Called H.2 receptor blockers, the drugs act by blocking the action of body-chemicals, histamines, which would otherwise under stress cause the release of excess stomach acid. The same idea may be used in the future for the treatment of other common illnesses caused by overstimulation of other parts of the body by histamines, including some forms of arthritis, and severe headaches. H.2 receptor blockers are also now being used to prevent internal bleeding after surgery.

Chemistry of the Brain

After a number of false starts, researchers are beginning to shed some light on abnormalities of the chemistry of the brain which may be responsible for some common mental illnesses. Thus work by Dr Leslie Iversen at Cambridge has shown that schizophrenia is often linked to abnormalities of one of the several neurotransmitters (chemicals which transmit messages between brain cells in the process of thought, or which modify general brain activity), the neurotransmitter dopamine. There are suggestions, from research now in progress, that drugs already used to treat heart conditions may have value in the treatment of schizophrenia.

The newly emerging understanding of how neurotransmitters work in the brain is also being put to work in treating brain diseases which cause not mental but muscular symptoms. Myoclonus, a brain disease which causes uncontrollable limb-jerking and tremors, is being successfully treated with the neurotransmitter serotonin. Overall, it begins to look as though a number of brain diseases, causing both mental and physical symptoms, may be due to deficiencies in the various chemical communications-systems between the brain cells. These deficiencies may be, at least partly, reparable from outside.

Senile dementia, with loss of ability to speak or read, aimless wandering, inability to concentrate and deep depression, is usually considered no more than an end-product or acceleration of the normal ageing process. Yet in recent years evidence has come to light which strongly suggests that senile dementia is quite a separate condition from old age, that it may be amenable to drug treatment, and that research to develop such drugs would be potentially perhaps the most cost-effective research programme ever undertaken, in view of the enormous amount of special care now required for many thousands of wretched elderly people suffering from the condition. Doctors at the London

Insitute of Psychiatry have shown that senile dementia is probably due to particular biochemical defects, missing or deficient brain chemicals, in turn perhaps due to an infection. This gives hope of artificially supplying the missing chemical, or eliminating the infection, which might avert the progressive deterioration which is at present inevitable.

Three Advances

Another exciting technique, now being pioneered in Britain, is that of replacing some organ transplants by injections of living solution. This technique looks so promising, on the small or experimental scale on which it has so far been tried, that it begins to look as though the hypodermic syringe may replace the scalpel for some important purposes.

Cells which produce sex hormones, insulin, adrenalin, and the body's immune response to disease have now all, experimentally at least, been transplanted, not in the conventional way in whole organs but by the much simpler and safer procedure of injecting them in a solution, by hypodermic syringe. In this way people with defective ductless glands or enzyme-deficiency diseases can have their deficiencies repaired without having to undergo surgery. Newcastle University's Department of Surgery and the Kennedy Institute of Rheumatology in London are among those pioneering this technique.

Burns are among the most painful and damaging consequences of accidents and assaults. The Burns Research Unit at Birmingham has developed and tested a vaccine which gives such quick protection against the bacteria which commonly infect burned tissue that it can be given to burned patients as an alternative protective treatment to antibiotics. This gets neatly round the fast-growing problem of emerging resistance to antibiotics on the part of the bacteria. Another problem in the treatment of burns is that skin grafted on from a donor is often rejected, being "foreign" material. Doctors working at Dundee University have found a way to treat grafted skin so that it is not rejected. Tissue treated in this way can be used to repair severe injuries and replace flesh and bone, as well as in skin-grafting.

Eye injuries are another especially distressing consequence of some accidents and assaults. Corneal grafts put in to restore sight after accidents have a poor record, compared to those inserted to repair damage due to eye disease. But at the McIndoe Memorial Unit at East Grinstead, Professor Richard Batchelor is building up the world's first bank of tissue-typed corneas, stored in deep freeze, each with its tissue constituents precisely known and labelled, so that each cornea can be matched precisely to its recipient. In this way Professor Batchelor has shown the prospects for corneal grafting after severe injury can be increased from almost certain failure after a few months to a 95 % chance of success.

The linked fields of contraception and fertility are seeing big advances too. Research workers at Cardiff are developing chemicals which will reduce the dose of the sometimes dangerous sex hormones in contraceptive pills several times over, without impairing efficiency. Roger Short of the Reproductive Biology Research Unit at Edinburgh is testing a technique whereby women can be harmlessly protected from conception for a year by a single injection.

At the opposite end of the scale, women who desperately want to give birth can now be aided by bromocriptine, a drug developed at St Bartholomew's Hospital, which stimulates fertility without the risk of multiple births. Dr Steptoe's work, which allows the fertilisation of human ova outside the body so as to bypass blocked or impaired Fallopian tubes, is of course world famous. Meanwhile Dr Robert Winston in London, with Belgian colleagues, has developed an alternative operation to transplant Fallopian tubes complete.

Helping the World

Medical research is a matter of international co-operation – though also often of intensive competition – and most of the research projects described here are interleaved with work in other countries. And so it seems right to end with a few examples of the real benefits that medical research in Britain is bringing to the countries most in need of it.

British workers are contributing largely to the development of a vaccine to protect against malaria, which should become available within a few years. They are taking the first real steps towards a vaccine to protect against sleeping sickness. They have developed and used a test to detect a foetus suffering from incurable anaemia – common not so much in Britain as in the Middle East and Asia. They have found a way to grow the leprosy bacillus outside the human body, first in the feet of mice, more recently in specially bred whole mice, so that new drugs against leprosy can at least be developed and tested without subjecting humans to unacceptable risks.

Further in the future, the genetic engineering and other techniques being developed at Cambridge in the Molecular Biology Laboratory, are showing the way towards huge living factories of cultures of cells and bacteria, devoted to the mass-production, on an unprecedented scale, of antibiotics and even of antibodies, the human body's own chemical weapons against disease.

Often nowadays we are made to feel that Britain is living on past glories. But every piece of research I have tried to describe in this all too short space is going on *now*. Medical research is often relatively inexpensive, its benefits are enormous, Britain's talents for it are unquestioned. Let us hope that, as we all tighten our belts, these projects and the many others like them which I have had no room to describe, do not suffer.

Prosthetic Man

The British Genius Exhibition displays several of the replacements for joints and other bits of human anatomy – known as prostheses – recently developed by British engineers and doctors. Such replacements can revolutionise the daily lives of crippled people, but making effective prostheses is more difficult than it looks.

The commonest site for arthritis is the hip joint, which bears heavy loads throughout life. In 1938 Philip Wiles of the Middlesex Hospital developed the first stainless steel total hip replacement. In 1951 it was re-designed by G. K. McKee in Norwich in the next major step. Thanks to them, the replacement of hip joints became an estab-

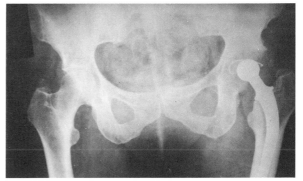

lished operation. In 1965 John Charnley of Wrightington Hospital, Wigan, improved the method of fixing the replacement ball-and-socket joint to the thigh bone, using a dental cement. But body fluids do not lubricate steel sufficiently. Charnley found a plastic which would stand up to the wear. He also cut infection of the wound, which used to be a serious hazard, by putting the surgical team in special suits from which their exhaled breath was withdrawn through a tube. Many thousands of patients have been helped back to normal movement by the pioneering work of these men.

Surgeons and engineers were also giving their attention to the knee, another common site of arthritis. The knee is a more difficult problem. The first designs, based on a hinge, were developed at the Royal National Orthopaedic Hospital at Stanmore. Later hingeless joints were evolved in which the bearing surfaces slide over one another,

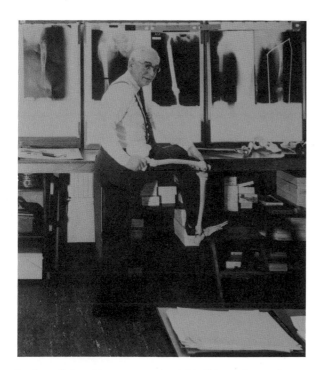

Professor Scales of the Biomedical Engineering Department at Stanmore shows a leg; behind him can be seen the X-rays of a total hip-femur-knee replacement

The total hip replacement can be clearly seen on this X-ray of the patient's pelvic region

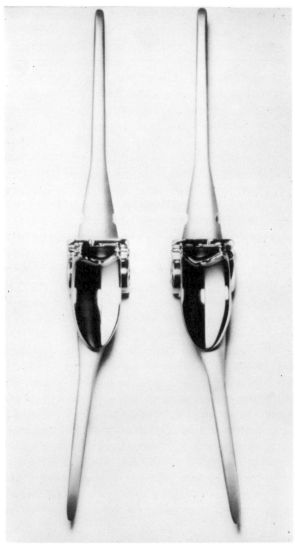

held in contact by the ligaments and natural elements of the leg. There are now many other versions: for example, the one developed by Professor Verna Wright has cappings modelled closely on the ends of natural bones, whereas Professor Swanson at Imperial College London prefers smooth surfaces.

After the knee, the even trickier problem of the shoulder: the arm moves in many directions, and the joint relies entirely for its stability on muscle control. The artificial joint must be linked to the shoulder blade, a rather curiously constructed piece of bone. The first total shoulder joint was developed in the Department of Biomedical Engineering at Stanmore; and a group at Leeds School of Medicine has devised another version which they believe will stand up to the most violent exercise anyone could reasonably take.

An arthritic ankle is exceptionally painful; standard surgical practice was to fuse the bones together. This meant three months in plaster and a limp ever after. It was Dr Kempson of Imperial College who devised an ankle prosthesis and Mr M. A. R. Freeman of the London Hospital who fitted it (he also fitted the Swanson knee mentioned above). The patient can walk within 48 hours, though this is not recommended as the surgery must be allowed to heal.

After all this, fingers and toes are receiving their due attention. There remains the elbow. At the Norfolk and Norwich Hospital and at Stanmore hinged elbow joints have been developed and been in use for a number of years. A re-designed version,

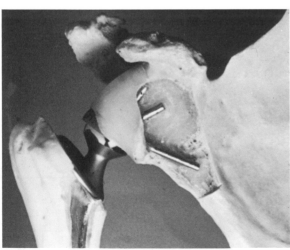

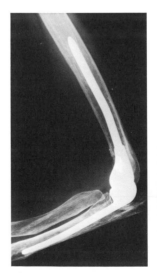

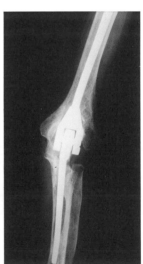

Left- and right-handed Stanmore Alivium alloy knee joints: the central flat plate is the kneecap-bearing area

The Stanmore shoulder joint is shown as it would be embedded in the bone of the shoulder-blade and humerus

Two X-rays of the same Stanmore hinged elbow joint in situ

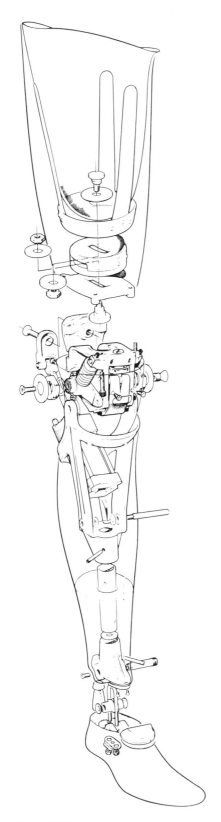

The Blatchford artificial leg

by Mr R. Dee, with sliding surfaces and snap fit is just coming on the market. Unless you want to replace a broken neck, that about completes the order.

Prosthesis of a more elaborate sort is required for patients who have lost a limb. To provide an effective artificial leg is not easy. The knee must move easily while walking, yet must remain stiff when kneeling or kicking. It takes some time to learn to manage an artificial leg of modern design and it is easy to fall over. The Blatchford leg, which won a Queen's Award for Industry, has been a world-wide success. Whereas in the past each leg had to be tailored to the needs (and height) of the individual and had to be replaced when he was ready to graduate to something more complicated, the Blatchford leg is built from modules and can be made more complex as the user progresses. It includes an artificial "tendon" the resistance of which can be adjusted as needed. In fact, one can even kick-start a motor-cycle with it!

Prosthetic help is also available for the deaf, and may one day be available for the blind. Ordinary deaf-aids amplify so much that they tire the wearer and may gradually deafen him. Professor Richard Gregory has devised a deaf-aid which avoids this by sampling the signal and rejecting unusual features. And Professor Brindley is working on a visual prosthesis for the blind. The patient wears a hat containing a miniature radio transmitter from which fine wires run to the surface of the brain. Flashes of light are seen, and shapes (such as letters and figures) can be discriminated. This is a long way from the restoration of vision, and involves surgical intervention to implant the wires, but it is perhaps a beginning. Meanwhile improved Braille-readers are making it possible for several thousand blind people to continue work and even to operate computers.

Arthritic Disease
Probably more people are affected by "arthritis" (i.e., diseases of the joints) and "low back pain" or "slipped disc" than any other degenerative disease. These are diseases of cartilage, the rubbery material which lines the bearing-surfaces of joints and separates the spinal vertebrae. Like rubber, it seems to lose its elasticity with age, and it collapses under repeated stress.

Diseases of cartilage affect at least 3 million

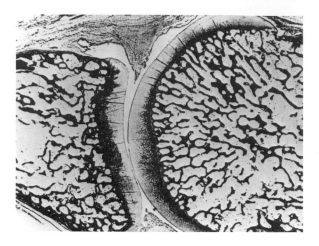

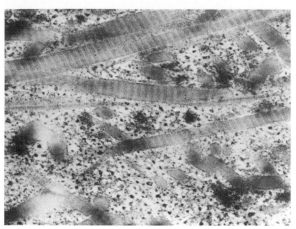

people in Britain, of whom 1 million are handicapped to the point where they cannot work or look after themselves properly. These are figures from a 1971 national survey; but some local studies suggest the proportion of afflicted people must be even higher. Almost two-thirds of those interviewed (of working age) claimed to have had symptoms, and one-quarter said they had been off work for this reason at one time or another. The loss of industrial effort, in 1971, was over 37 million days, far more than was lost in industrial disputes. Backache alone accounts for 7 million days and cost the country £100 million.

Research

What are we doing about this? Some people would say: not enough. It is not lack of money but lack of staff and methods of attack which are the obstacles.

Studies of the mechanical stresses on joints show they bear three or four times the body-weight at the moment the toe leaves the ground in walking. The pad of cartilage distributes the weight over the head of the bone. If the cartilage gives way, the weight becomes concentrated at a single point, and the bone begins to wear. The joint becomes painful.

Substantial progress has been made by British researchers who have found that cartilage consists of spongy lumps (proteoglycans) held in a framework of rods (collagen). When a joint takes a load, water is slowly squeezed out of the "sponges" and later re-imbibed. Normal cartilage holds 70% water; damaged cartilage holds even more, hence its softness.

Dr Helen Muir has shown that the proteoglycans form long threads. It seems to be these which give way in arthritic joints. But there are changes in the collagen too. Research continues.

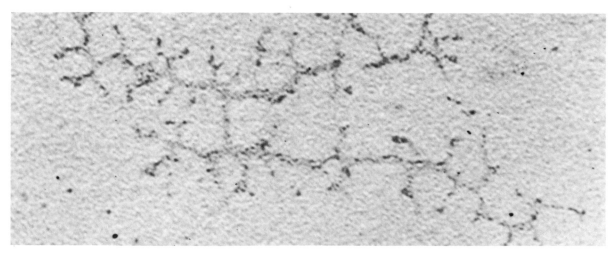

Section of a normal finger joint, showing the cartilage pads

Electron-microscope photograph of proteoglycan threads

Section of normal articular cartilage, seen under the electron microscope, showing collagen and proteoglycan

How Healthy is the National Health Service?

While the British National Health Service is in many respects a model, it is becoming alarmingly costly and clumsy. It costs £5,000m. a year, and the cost is steadily rising. It is the largest employer in the country and the tenth largest employer in the world. There is one bureaucrat for every $5\frac{1}{2}$ patients. It may be a source of legitimate pride, but it bears thinking about.

When Lord Beveridge proposed the idea, he imagined, with sublime optimism, that costs would steadily fall! He argued that a health service would make people healthier, so there would be less to do. In actual fact, if people live longer they survive to have the chronic diseases of old age which do not kill but call for prolonged care. Besides this, when treatment is free, some people come for help they do not really need, or for reassurance, or to justify a day off work. The Health Service has also taken on jobs, like abortion and birth control, which Beveridge did not contemplate.

But the biggest problem is the changing age-structure of the British population. We have 9.3 million people over retiring age (29 % of them are over 75 and this will rise to 37 % or 3.5 million by the end of the century). Already some 42 % of the population are either over retiring age or under school-leaving age – and have to be supported by the other 58 %. Thus the medical burden is heavy and will grow heavier.

Some people argue that we should spend more time preventing illness and less in curing it. Easier said than done. Many serious conditions are linked to personal habits – a high fat diet and too little exercise with heart disease; alcholism with drinking, etc. Motor accidents are another grave burden, taking up 3,000 hospital beds costing the N.H.S. £40m. a year, not counting £60m. in other costs to the country. And if people increase their chance of arthritis by violent exercise, how is one to stop them? Dr David Owen, until lately junior minister of state for health, believes that people should be dragooned into caring for their health, now that failure to do so creates a national burden. (The legislation making seat-belts compulsory is a first step in this direction.)

Hospitals and Health Centres

Critics claim that the health service is too "hospital oriented". It is true that hospitals take about 40 % of all the money spent by the health service. Current policy is to get people out of hospital and into their own homes, but this is more to reduce costs than because there is any positive concept of a total health-care system.

Before the war, in Peckham, an experiment in total health care was tried: a centre was built to which people came for sport and social reasons as well as health care. Their state of health could be judged in relation to their total pattern of life and efforts could be made to get them to change lifestyles which were causing psychosomatic disease or undermining their resistance. This centre was closed down when the N.H.S. was inaugurated, on the grounds that unless everyone could have such care no one should. Maybe we should opt for more health centres and fewer hospitals. The hospital community centre which has been designed for Cannock, by Denys Lasdun of National Theatre fame, is a step in the right direction. Planned for easy access, with special provision for the elderly and their needs, doctors can see their patients with a wide range of technical facilities at their disposal. Alas, construction has been indefinitely postponed.

We need to abolish the artificial distinction between health and welfare and to distinguish the chronically sick from the healthy aged. We could provide more home help. We could have day-hospitals, with diagnostic and curative equipment, but no beds, for the "walking wounded". We should make separate provision for the mentally sick who occupy beds in hospitals organised to deal with physical illness.

A massive re-organisation of the health service is one of the problems lying ahead. We should decentralise control: at present there are 90 area boards answerable to 14 regional boards in turn answerable to the Department of Health and Social Security. In the last two tiers alone there is a staff of over 12,000! Dr David Owen argues that this top hamper is necessary, and even that it needs strengthening, in order to ensure equality of care in the different parts of the country. (No one knows how such equality is to be determined, given variations in climate, age structure, jobs and so forth.) The diversion to actual health care of the £2,300m. paid in salaries to these officials might do more to raise standards than committee-meetings and paper-pushing, I suggest.

Challenge No. Five
No Place Like Home

The home is a British conception: the word does not exist in French or German. The notion of a cosy fireside where one is accepted and at ease, which one might suppose to be universal, is not so. The very word "cosy" is probably derived from the French "causer" meaning to have a conversation, and has clearly changed its connotation a good deal in crossing the channel.

Having invented the home, British Genius turned to the garden, which it conceived as a series of outdoor rooms, delighting the eye but with conveniently placed seats where one might read, walk, flirt, or simply snooze. And from the garden we pass almost imperceptibly – thanks to that genial eighteenth-century invention the "ha-ha" (an unobtrusive ditch, so called because people said "ha ha" when you fell into it without warning) – to the landscape at large. Only France, I think, can claim to have produced landscape gardeners at all comparable with "Capability" Brown and Humphrey Repton; but the French approach was more formal. The idea of landscape as nature beautified was a British inspiration.

So it is by a natural progression that one of Britain's postwar achievements, in which she has led the world, has been in the field of town and country planning. Though not without defects, in the eyes of conservationists, it has preserved us from much of the spoliation one can see in other countries.

After the war, we were the first to build New Towns – not mere building estates, but conscious attempts to foster a workable social structure and a sense of urbanity.

We made some mistakes but we avoided others, and people from all over the world came to look at them. Today we know more about how to involve people in planning their own environment, and developments like Newcastle's Biker – criticised in some quarters for its casual rough and readiness – may prove to be models which others will be glad to follow.

Gear Change
Since the war British creativeness has been much occupied with matters domestic. British styles in gear for the young – the miniskirt and Carnaby Street will serve as symbols of something actually much larger – affected the whole civilised world. Again, British toys and nursery items are sold world-wide and model-making is one of our best export lines.

Although Voltaire called England the country of a thousand religions and only one sauce, we have actually created as wide a range of local dishes, relying not on sauces but intrinsic excellence, as anywhere in the world: from haggis to Chelsea buns, from Edinburgh rock to Yorkshire pudding. And if, between the wars, the food in popular eating places was often dull or inferior, more recently there has been a revolution in British cooking, eating and drinking. More people take more trouble at home, while the number of restaurants serving food of high standard has multiplied out of all knowledge. We have even instituted a wine-growing industry – some of the British white wines stand comparison with German equivalents – while British beer – surely a product of genius and certainly quite different from Continental beers – is winning customers despite the desire of Common Market countries to substitute their own insipid brews.

Last but not least, we can pride ourselves on Britain's achievements in the field of education. Not only have we built, since the war, numerous new schools of imaginative design, but we have extended education to more people for longer periods. Most originally, we have opened a second chance to those who missed out when young, by founding the Open University. Launched to gloomy forebodings in many quarters, it is now beginning to be copied in other countries.

House and Home
But what of the home itself, the actual structure? The great problem for the future to solve is the provision of adequate housing for everyone. On the basis of surveys made in 1971, and allowing for what has been done since, it looks as if we have at least 1,500,000 houses which are "substandard" in the sense that they do not have indoor sanitation, a hot-water supply and other basic amenities. But this is not the whole extent of the problem, for there are many people who are homeless, "squatting", living in caravans or boats from necessity rather

than choice, or living in crowded conditions – including married couples obliged to live with parents.

The rate at which we are tackling this problem is not impressive. We build some 300,000 new houses a year – but some of the houses which are thus replaced are not technically substandard, so that the net gain is less. Many of the newer building techniques intended to speed up rehousing are only effective when applied to large blocks of flats or repetitive designs, and today we realise how many disadvantages these structures have for their inhabitants. Geniuses are wanted to devise more rapid and flexible methods.

A recent international study showed that the British building industry compared unfavourably with those in the Continent in productivity. To rehouse Britain properly means solving this problem too.

But, once you have a house, the most pressing problem today is paying the heating bills. The house of the future is the low-energy house, and here British ingenuity is definitely at work. I have therefore asked Julian Keable, an architect who has specialised in energy-saving in existing houses, to give an account of progress and prospects.

Warmly Recommended

by Julian Keable

Although we have built a huge number of new homes since the war, the fact remains that only about half our housing stock has been replaced in that thirty-year period. Moreover, the rate of new building has been steadily dwindling, and it is obvious that we are unlikely to be able to increase the rate for some time to come. Even if we could keep to the level of 300,000 houses a year, by the year 2000 we should still only have replaced some 26 % of the total and in fact it is much more likely that only some 10 to 15 % of housing will be replaced by that time. Therefore, the house of tomorrow is the house you are living in now! You may be delighted with it, or you may be appalled by this thought, but either way the problems of the house of the future will have to be solved by changing what we have now.

Most of us are already worried about the cost of the heating bill – both for hot water and room-heating. In many cases the amount of space we have needs changing in some way: either we need more, or less, or space of a different kind. If we own our own home then the question of money will be important; both what it is costing us to buy, and what it costs us to run and to repair. If we are renting then it comes to the same thing: somebody

A house adapted for self-sufficiency: the flat roof on the left is for the collection of rain-water, and there are solar panels on the central roof

else may be responsible but the problems are there to be solved.

We seem to have decided pretty firmly that we want to live in a house rather than in blocks of flats and I shall concentrate my attention on this field: particularly again since most of the existing stock is indeed in the form of houses.

While there are still many houses which are past saving and simply have to be cleared away, it is being realised more and more clearly that there is a great deal of value in a house if it is still sound: in many cases we simply could not afford today to build to the sort of standards which were at one time commonplace. Even to pay for the timber in the joists or the brickwork in the walls would put many of the older houses right out of reach in terms of new construction.

Naturally, most of the experimental and forward-looking work is done in the context of new housing, but the ideas that are developed and the products and services that develop out of these in turn will be of the most use in direct proportion to their applicability to our old houses now.

Would It Pay?

In Britain less money is being spent on alternative energy research of all kinds – and particularly in the field of housing – than almost anywhere in the world. (Even India, poor as she is, can be proud of spending more on energy research than we do!) Nevertheless work of all kinds is going on in universities, by individuals, by government departments and local authorities and within industry and commercial concerns. In many instances it is not the cost of sitting down to think an idea out or even the cost of putting that idea into practice on an experimental scale that is the problem.

Take, for example, the idea of heating water by using the sunshine. On the face of it, provided it works it must be worth while since the sun is free. Assuming you can afford the outlay, surely you should? A carefully monitored programme of research in this case would take into account, however, such things as the amount of hot water produced at different times of year; the effect of location (if you live in an area of high pollution less sunshine will arrive); how long the hardware is likely to last before it needs replacement; an evaluation of what you would have spent heating

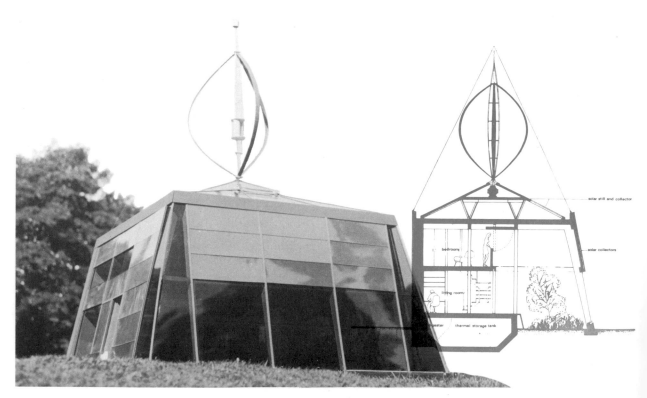

An exterior view of the Autarkic House from the south-west

A section drawing of the Autarkic House designed by Alexander Pike water by electricity or gas; what is likely to happen to these prices in the future and so on and so on.

Typical of such research is the work of David Sharpley of Don Engineering. With painstaking patience he has heated water in a row of cylinders each day for a couple of years using both solar collectors of his own design and also others he has bought in, trying out new materials, new controls. He was one of the first in the field and set out initially to convince himself, afterwards to convince others. Much the same is true of dozens if not hundreds of colleges, universities and research departments; hardly one that has not got something being tested up on the roof.

The Autarkic House

Shown in the exhibition is the Autarkic House – an extreme solution to some of the problems facing us. Self-sufficiency requires an entirely new building and this design has been particularly developed for use on isolated sites. Cambridge's Alexander Pike set out to discover whether it would be possible to build a house that needed no connections whatever to existing service grids or pipes; whether one could maintain the high standards now being set for urban living without an electricity pole, sewer or water-pipe anywhere around. One of the starting points of course, was to discover just how much we do pay for being connected to these networks: the results were astonishing. It was found that for a house situated near service networks (water, drainage, electricity, etc.) the cost of connection would not exceed £670 at most; but for a house a kilometre away it could cost as much as £6,000!

Next it was necessary to quantify in great detail just how much we need of the major incoming commodities of energy and water, and whether the natural elements of rain, sun and wind could provide these. At first sight the task seems impossible, but by thinking hard what each input was really *for* (for example, only a tiny proportion of the water we use is actually consumed in food and drink) and by rigorously avoiding waste at all points, a theoretical solution has now been arrived at, studied and checked by computer programme, which indicates that a design along the lines shown would work in our climate. As an experiment a Pike house is being built at Cambridge.

The next task is to solve all the detailed problems thrown up if you decide to collect all the rainwater falling on a house, and not to throw away the water from the bath and basin without having extracted its heat and perhaps purified it sufficiently for use down the lavatory. Finally the waste-products of the house can be processed so as to extract from them again methane gas, hopefully in sufficient quantities to do the cooking. That in turn requires the redesign of the stove and the oven so that they would not use nearly as much gas as we are used to. (Though methane from sewage is a workable idea on farms, where animal wastes are available, you cannot get worthwhile quantities from the human wastes of an ordinary house unless you add straw, shredded paper or other vegetable matter.)

Typical of this approach is the solution adopted for heating the house. Half of it would not be heated at all! That half would be a useful addition to the basic space, whenever the weather was mild enough, but only an inner core would be kept at a comfortable temperature in winter. Many of us adopt a similar plan in fact – but perhaps not so many would set out to arrange a house in this way. It certainly makes sense though, when fuel bills have already doubled over the space of a few years and may very well do so again.

If this approach is proved successful, Pike foresees a new tool in our hands when planning our land resources. Many a spot which already has a country road could have such a house built on it rather than trying to pack the same house into a village or town. The savings in servicing and drainage could contribute to the upkeep of a road which is at the moment under-used. With such houses, it would be possible to develop agriculturally-based industries and to offer similar housing standards to people in remote areas as can now only be provided at great cost or not at all.

Let us now take a look at what is being done elsewhere. The concentration on energy issues apparent in the Autarkic House did not happen by chance, of course – since if we are not already paying as much for our energy bills as we are in rent and rates, we very soon shall be: at any rate if we wish to keep warm.

Teacosy House

Another approach to energy-saving is that of architect Peter Bond. The firm of Wates has built in Wales, for the National Centre of Alternative Technology at Machynlleth, what is probably the world's most extreme instance of energy conservation applied to a house of basically conventional design. Bond refused to consider methods of collecting or supplying energy to his house until he had found ways of reducing that demand to an absolute minimum. The result is a house where the 50mm cavity (2in. if you prefer) of ordinary walls has been enlarged to 450mm (18in.) entirely filled with fibreglass: a teacosy if ever there was one. Not content with double-glazing the windows, he fitted two windows in each opening: four layers of glass. The same for the front-door; two front-doors.

Bear in mind that each of us gives off energy in the form of heat, and that the same thing is true of the cooker and the fridge (!) and the electric lights and the T.V., and that when the sun shines through the windows the heat cannot get out again in quite the same way. The "incidental gains" within such a well-insulated shell result in a house that needs only 20 % of the amount of energy that an ordinary house built to today's standard would require. Wates calculate that the extra cost of doing all this (and hence the extra mortgage payments) would be balanced by the fuel-saving taken over three to five years.

The small amount of heat still required is supplied by a heat pump: a device fitted inside every refrigerator, though usually it is the cold end which is of interest. In this case both heating and

cooling can be obtained by the use of a reversing valve in the system, the energy (electricity) being applied to a compressor and a fan which in turn

The first low-energy house built
by Wates in Wales

A heat pump in position,
revealed by cutting away its
lagging on the near side

extracts the heat from surrounding air (even in winter) and "pumps it up" to living-room temperatures.

A very different solution has been arrived at from a fairly similar starting-point by Robert and Brenda Vale, architects who at one time were working with Pike on the Autarkic House in Cambridge. They tend to place more emphasis on complete self-sufficiency including the question of food production and have not only written an interesting book on *The Autonomous House, design and planning for Self-Sufficiency*, but have put their ideas into practice in their own house near Ely. Having again reduced their needs by thorough insulation and draught-sealing they rely for the bulk of their additional energy-needs on a tall and elegant wind-generator of sturdy manufacture. Capable of producing as much as 5 kW at full load, it is cunningly devised to charge 12-volt batteries for all their lighting requirements, then, as wind-power increases, to heat standard electric off-peak radiators modified to the appropriate voltage. Incidentally when they do want to sit round the fire they burn waste sawdust in a specially designed wood-burning stove.

Town House Measures

All the houses mentioned so far either are, or are intended to be, situated in the depths of the country. What about houses situated, as most are, in towns? I have given this a good deal of thought and in fact put the results into practice.

My house is a typical London semi-detached with solid walls, draughty windows, an attic which is cold in winter and hot in summer, while part of it was a cellar only 4 feet high and therefore useless as a room. Arithmetic shows that thorough-going insulation is always worth while, not only in the roof (4in. of fibreglass in my case) but all the external walls as well. In the case of the fairly blank side-walls the insulation went on the outside so that the "thermal mass" would be retained (that is, the ability of the whole structure to heat up and conversely not to become cold rapidly). The other walls I lined internally. The draughty windows were dealt with by D-I-Y double-glazing and draught-strips; a door lobby and blocking of the remaining fireplace-flues cut down ventilation-rates drastically.

As a result of this the house needed far less heat –

but still a sizable amount. Off-peak storage heaters were using alarming numbers of units. After doing some theoretical work with a heat-pump engineer (Chris Dodson) we decided to go ahead with a standard American package unit. All the equipment was in fact "off the shelf" and rapidly installed. Heat is now collected from the outside air after first passing it on the underside of the slated roof where any solar energy available is picked up, and is calculated to account for something like 10% of the total energy requirement: all for the cost of a roll of building paper! The heat pump itself then supplies energy to a thermal store in the cellars: 2,600 gallons of water entirely enclosed in a plastic liner. This means that when the heat pump is not running (i.e., at night or during especially cold weather) the radiators continue to be fed.

A second example of this principle has now been embodied in a new house in Hampshire, only in this case the heat withdrawn from the thermal store is circulated to the house by ducted air and not by water in radiators.

Further Outlook

Let us look further ahead and assume that firms like Wesco develop useful and efficient wind-power machines, that firms like Farm Gas develop more widely applicable methane generators, and that appropriate technology is developed to deal with reducing water usage, collecting rainwater and all the other aspects of the Autarkic House shown in the Exhibition. Operating and servicing the hardware may daunt some householders. Some of it may in fact work more efficiently if it is a little larger than would be appropriate for a single house. For example, even a house that has reduced its ventilation rate to one air-change each hour is throwing away a great deal of expensively heated air. There is a device known as a heat-recovery wheel which could reclaim 75% of this – but the smallest one available would be sufficient to deal with three normal houses! It consists of a slowly-revolving wheel of high thermal mass, half of which is situated in the hot-air duct, and half in the cold. So it picks up heat from the first and gives it up in the latter. It may well be that advantages could be gained in pursuing a "grouped servicing" concept which would develop autonomy on the scale of 15 to 30 houses. This would permit the hardware to be "janitor"-operated and could avoid the need to get your hands dirty.

For a methane plant to work efficiently, it needs to be fed a certain amount of cellulose: straw or even shredded paper will do. Obviously, you cannot do this if you are out at work, or ill, or away for the weekend. Solar collectors need maintenance too if they are to work at optimum efficiency and there are a number of other things which sound sensible in bigger lots than that of the small back garden: keeping chickens, recycling refuse through compost, and of course, slightly larger windmills, which certainly could not be put on the roof of a house and certainly need an eye kept on them. So a group set-up would get round this problem.

Lifestyle

Our lifestyles have changed tremendously in the last couple of generations and look as though they will continue to do so. With increased mobility, different holiday habits, different working hours and different entertainments this is of course, inevitable. The Do-It-Yourself movement is an interesting example of the way in which circumstances have combined in the home. Shorter working hours, the rising costs of decorating and repairs and the development of relatively sophisticated home tools have come together to revolutionise the idea of the home handyman.

Our eating habits have changed enormously also: at one extreme people are increasingly interested in natural and health-food products, in making their own beer, baking their own bread and growing their own vegetables. At the other extreme, under pressure of time and circumstances, packed and prepared food of all kinds now being available, the very pots and pans themselves are changing into plug-in hot-pots and plug-in frying-pans with capacious lids, capable of cooking an entire meal. Thought has been given to updating the haybox and such a device might well replace the oven quite shortly: any oven well enough insulated would not require continuous heating.

Lifestyle is of vital importance when it comes to understanding the use to which we put our homes. Even such basic things as the pattern of people's use in opening windows and doors is something about which little is known, and so the G.L.C. undertook to discover the facts in their estate at Harold Hill. No good putting in a sophisticated ventilation system if people insist on propping the back-door open so as to let the cat and dog in and out.

Space

Obviously the changing size of our basic family unit is the largest single element to be taken into account here. The typical family with its one or two children and the fact that so few of us live together with our older or more distant relatives in the way that was so common in the past means that our typical dwelling unit is much smaller. It is therefore, in a way less flexible too, since it is impossible to afford separate rooms for all the separate activities we may wish to indulge in from time to time.

One result of this, of course, has been the need to provide far more specialised accommodation elsewhere – from nursing homes and geriatric units to mental institutions – in large measure filled by people who have faced loneliness or maladjustment of one kind or another. Many architects have proposed houses that could "expand" and "contract" to meet different times for the same family. Maybe many of our "too large" old houses will be partly used for work and education too. Possibly the development of communication methods and techniques both in the fields of education and business will lead towards a greater use of our homes as places of education and business: our own children's education and our own livelihoods. Why provide an office for a salesman or a service engineer who is going to spend most of his time seeing his customers?

It might be more useful to have a "home office" – a study – linked by telex to headquarters, and so save time and fuel by avoiding the need to pick up instructions. Why provide a classroom for that part of the child's education which consists of just plain learning by heart: perhaps spending more time at home could have some advantages? For many children, videotape cassettes would allow them to go at their own pace, in congenial surroundings, with the chance of much more personal instruction at school.

In short, the challenge of the next twenty-five years is to discover what changes in the design and location of our homes are called for as a consequence of the dramatic changes which are occurring in the way we live – changes which spring from having more money, more leisure, more education, wider aspirations and a greater expectation of life.

Five
Genius Wanted!

Genius consists in seeing what everyone has seen and thinking what no one has thought. *Albert Szent-Györgyi*, the discoverer of vitamin C.

The solar eyeball, the sun-seeker device being developed for use in the direct conversion of solar energy to electrical power: using energy from the sun and a new type of magnetic drive, the eyeball keeps itself in alignment with the sun.

Half a century ago, Sir Robert Hadfield, at the request of the Institute of Patentees and Inventors, noted as inventions which were wanted: a cinema film which would speak; a method of utilising atomic energy; a process for instantaneous colour photography; a method of conveying speech direct and readily to paper; a means of regulating rainfall; the secret of instant memory, such as our brains achieve, without running through a card index. He himself suggested four: a way of seeing through fog; a metal twice as strong as any known without brittleness; a non-corroding, heat-resisting furnace lining; and a method of stabilising ships.

Ninety per cent of these wanted inventions we now have.

But, as the preceding essays have made clear, there are many areas where ideas of genius are urgently wanted. One obvious need is a much better way of disposing of or neutralising radio-active wastes. Yet the National Research Development Corporation reports a shortage of ideas worth backing. In 1975–76 inventors sent in 1,854 ideas, asking for support, but only 157 of these were found worth backing, and the Corporation's previous experience suggests that only 21 of these will finally make a success. (Say $8\frac{1}{2}$ %; for private inventors the rate is only $2\frac{1}{2}$ %.) So the Corporation is nowhere near the limit of its funds: it has power to borrow up to £50m. if necessary. And it claims that there is no known example of an idea which it rejected which subsequently went on to succeed.

In deciding what to invent, a good deal depends on what you regard as better. Is it better to invent a faster aeroplane or a safer one? Should one concentrate on the problems of the under-developed countries or simply go for the markets you know? Professor M. W. Thring, of Queen Mary College London, has recently, with Professor Eric Laithwaite, written a book called *How to Invent*, from which, with kind permission, I have culled the following suggestions.

1 Things to help the Third World: a cheap, effective irrigation system; a better way of making leaf protein, low-cost buildings for crop-storage; cheap ways of transporting equipment across rough or roadless terrain.

2 Things to do dangerous work: remote-controlled robots could do coal-mining, cap oil wells, enter

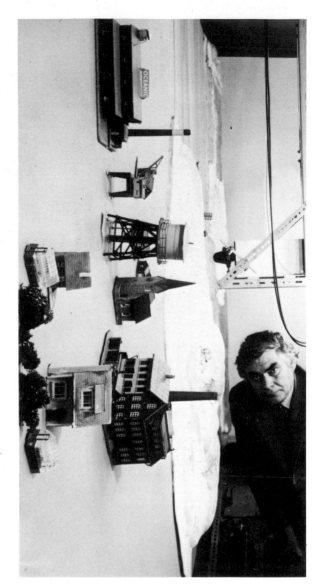

burning buildings or rescue people from crashed aircraft, etc.

3 Medical engineering: help for elderly, blind, arthritic or limbless people; sewing machines for surgeons, to replace the laborious hand-sewing and knotting by which damaged structures are mended; help for nurses (who are often strained by lifting patients); better ways of distributing hot food and medical supplies.

4 Making work more interesting: machines which will do the dull jobs of making, packing, testing and shifting goods about in the factory.

5 Education and communication: portable memories; portable translation machines; dic-

Professor Richard Gregory displays his simulated landscape for the training of air-pilots: the camera can move over his Toytown to simulate almost any aerial approach

ORACLE, the
television-displayed news
service

An air-cushion transporter at
work in Alaska; normal
conditions include moving
across pressure ice, ice-floes, and
flood water

tation typewriters; access to news on television screens.

6 Travel and transport: improved goods handling on railways; ways of raising boats on canals from one level to another to replace cumbersome locks; cross-country transport by airships, distribution of newspapers, laundry, milk and daily needs by pneumatic tube; conference video-telephone system.

Many of these suggestions are already engaging inventors, and examples of some of them can be seen in the British Genius Exhibition.

One of the most far-reaching inventions would be an improved electric battery, which would make electric vehicles as flexible and efficient as petrol vehicles and so ease the oil shortage. But the really earth-shaking inventions are more than mere improvements, they open up needs we did not even know we had. Radical or not, the inventor must resign himself to meeting the following three objections:

1 It will not work.
2 If it does work, it will not pay.
3 Someone else thought of it first.
But that's life for you.

The Patscentre electric bicycle

Six
What Price Genius?

Genius is one percent inspiration and ninety-nine percent perspiration.
Thomas Alva Edison, misquoting Emerson

Genius, which means the transcendent capacity of taking trouble, first of all.
Thomas Carlyle

The McClaren baby buggy
shows its strength: this simple,
light and cheap pushchair has
eased the lot of modern parents

To take pride in achievement is all very well, but can easily develop into complacency. To the bland diet of self-congratulation it is proper to add the condiment of self-criticism if only to avoid the sensation of nausea. We may be a creative people, but we are also in dire economic straits, and it is natural to ask why our inventiveness has not prevented this situation arising. Two questions suggest themselves: do we in fact offer the most favourable opportunities for creative talent? and do we exploit new ideas with the necessary drive and determination?

At the time of the Napoleonic wars, in 1816, Francis Ronalds offered an electric telegraph to the Admiralty, who replied that they saw no need for any form of telegraph other than the one they had, which was a system of movable arms on hill-tops. A chain of such relay stations stretched from London to Deal and to Plymouth. Sixty years later, when they were offered the steam turbine, the Chief Naval Designer declared that any ship propelled from the rear would be unsteerable – this despite a demonstration in which their Lordships were successfully taken to Limehouse and back. In fact, the chief naval designer was against metal ships and steam power generally. The Army was no better. When Parsons offered it a shell which would carry a bigger charge further, because it spun in flight, they were uninterested and he was obliged to take it to the French. Not only practical inventions but scientific ideas are often met with unmerited scepticism. When Thomson announced that the atom was composed of smaller particles, this was ridiculed in many quarters. Since the word atom

meant indivisible, how could it be divisible? Even today, when we have learned that no idea in physics, however improbable it may seem, can be dismissed, we remain incredulous in other areas. The idea that cancer might, at least in some cases, be caused by a virus was treated as nonsense as recently as the 1960s. Just prior to that the Astronomer Royal declared that the idea of space-flight was "bunk", while a long paper in *Nature* had proved, to the author's satisfaction, that it was impossible ever to place an object in orbit round the earth.

Scientists themselves are sometimes the worst offenders. Lord Kelvin, who as we have seen was ingenious in the extreme, could see no practical use for Hertzian waves (radio) except possibly for communicating with lightships, while Lord Rutherford thought there would never be a practical application of the energy locked in the atom, and frequently accused those who thought otherwise of sensationalism.

Perhaps lack of imagination is one of the penalties of our practical British nature.

It'll Never Replace the Horse

But what maddeningly obstructs the inventive genius is the argument that, even if it works, it will never be practical. After Babbage had devoted much of his private fortune to demonstrating the possibility of building a calculating machine, (incidentally helping to create the machine-tool industry by his demand for more accurate cog-wheels), the public judgment was that it would forever remain "only a theoretical possibility".

It would be nice to think that today, fed as we are on science fiction and scarcely less astonishing science fact, we were more open to new ideas. But Whittle had to struggle from 1926 until 1934 to get support for his jet engine. Cockerell fared only a little better with his hovercraft. And we failed to exploit the potentialities of penicillin for more than ten years.

I fear that today such stories are in danger of being repeated. Take, for example, the super-conducting motor. This depends on the fact that at temperatures near absolute zero an electric current encounters no resistance and will flow for ever. (Some energy has to be expended of course in maintaining the low temperature around the wire or magnet.) One application of this is a vehicle

The flexible Dracone, backed by the N.R.D.C., is used to transport oil, water or other liquids

which floats above a metal track, by virtue of super-cooled electromagnets embedded in the floor. British authorities have rejected this on the grounds that it would mean laying new track – and anyway who wants to travel at 300 m.p.h.? Meanwhile the Japanese have built a prototype which successfully carries four people, and have put up £120m. to build a full-scale passenger service which will be operating in the 1980s.

Even if one does not want to travel at 300 m.p.h., super-conducting motors offer great advantages for ship propulsion, producing more horsepower for about one quarter the weight.

A British firm, International Research and Development, built a prototype, which was tested at Fawley Refinery, where it drove a pump, but the Central Electricity Generating Board was not interested. "There is no reason for pressing on with the development of super-conducting motors as a matter of urgency," it announced. But Dr Appleton, of I.R.D., whose baby it is, declares: "We are ready to market super-conducting motors now. We could fit them, for instance, in a steel-mill." And he ruefully adds: "Having done so much early work we are now in danger of falling behind." The sad fact is, electrical engineers have simply not yet woken up to the possibilities. Super-conducting magnets have many possible applications. An obvious one is to separate minerals or extract metallic pollutants from sewage. Dr E. Cohen, of Imperial College's Department of Mining and Mineral Technology, says we could produce a prototype in eighteen months and be in full production within five years.

If buyers are unenterprising about inventions which are ready for use, one can understand that they are even more lackadaisical about inventions which still require development. Thus the Coal Board is apathetic about attempts to design remote-control mining-devices, which will eventually enable miners to work in a normal environment, and which have an especial advantage that they can work in dangerous atmospheres or risky conditions where a human worker's life or health would be at risk. The Board does not expect such devices to be practical in the next twenty years or so and seems unwilling to become involved with the difficulties of a new technology before it is obliged to.

Another area where we have done the spade-work, but everyone hesitates to gather the crop, is in

the field of carbon fibres and similar reinforced materials.

Complacent Monoliths
Both the inventor and the maker of a new invention find themselves in a difficult position when the State is the only customer – that is, when the purchaser is a nationalised industry, the Health Service, the Armed Forces, the B.B.C., etc. If the monopolistic buyer turns your invention down, there is nowhere else you can take it. If the monopoly likes the idea, it can impose conditions, compel you to alter the design, fix an arbitrary price – and then, when you have invested in manufacturing resources, change its mind and pull out. Thus the makers of ambulatory monitoring devices (see p. 128) prefer to sell them to American and other hospitals rather than to the Health Service. The Post Office has lagged behind technologically and has repeatedly been criticised by the suppliers of equipment for chopping and changing. The trouble with a large organisation is that it makes large mistakes and tries to cover them up. If a dozen small organisations try out a dozen ideas, it soon emerges which is best.

Large organisations are given to complacency, and come to feel, quite genuinely, that they are efficient. Obviously, if they thought they were not, they would take steps to correct things. But independent observers may take a different view of what is efficient.

Britain, I believe, is unlikely to lead in innovation while organised in large, quasi-monopolistic units, whether publicly or privately owned.

Appleton's prototype
super-conducting motor drives
a pump

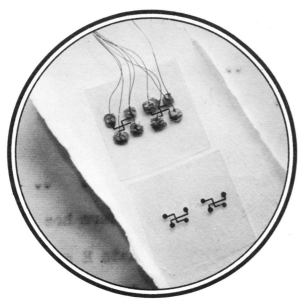

Lack of Enterprise

While some inventions languish because people cannot see the possiblilities others are left on the shelf because no one wants to spend the money necessary to develop the idea into marketable form – to "iron the bugs out". With complicated modern devices it can take millions of pounds to carry an invention through the stages of development. (It took £4m. to develop float glass, as Tom Margerison records.) Too often in Britain, we have taken the line: let someone else do the development work and if it turns out to be any good, we will acquire licences or try and break the patents.

Hesitancy in developing a new idea is no novelty. When Swan invented a successful light bulb it was Edison who came over and started building the power-stations on which it depended.

Inventors have repeatedly had to use up their private means, often reducing themselves to penury, before their ideas were accepted. Bessemer was only able to introduce his steel-making process thanks to the fortune he had made selling gold paint. The modern inventor is so poorly rewarded and the patent laws are so unfavourable to originality, that he has little inducement to struggle in this way. As H.R.H. The Duke of Edinburgh said in a recent speech: "The successful technological innovators must become heroes again. They must be allowed to gain and keep their rewards for success . . . If the rewards for success are adequate, many will take a chance when gambling on racing or football is treated more generously than investment in the wealth and employment-creating ideas of engineers."

The Josephson junction is a very sensitive electronic device with many uses: e.g., it can detect very small voltages and small magnetic fields; or it can be used as a high-speed switch in a computer

The bentonite soft-earth tunneller, which has been called the greatest advance in tunnelling since the Greathead shield of the 1880s, was developed with the backing of the N.R.D.C.

For years now, British patent laws have been the subject of bitter criticism.

Our patent laws only provide sixteen years' monopoly – but it often takes longer than this to get an idea off the ground. The jet engine was originally patented in 1867; Stringfellow's airplane in 1842; float glass in 1902. And one can only patent a mechanism, not an idea. While this eliminates mad ideas, like the man who proposed to freeze the clouds and mount guns on them during the war – it also means that anyone who can find a slightly different way to carry out an idea can obtain a patent for his mechanism.

The National Research Development Corporation was established to help inventors find the capital for development, and has financed some important inventions, notably the hovercraft and the cephalosporin antibiotics. But it is also accused of playing safe – it has never used more than a fraction of the £50m. made available to it and at present is earning more from its patents than it is paying out to develop new ones. And sometimes it takes a very short view, as when it recently sold off Hovermarine to America at a bargain price.

Currently, a factor which restricts inventiveness is the government policy of supporting research for which there is a clear practical application, and eschewing long-term and fundamental enquiry. But many important advances, perhaps the *most*

important, come incidentally to other work. Hertz was not trying to invent a new means of communication when he experimented on Hertzian waves; Fleming was not looking for a new drug when he discovered penicillin; Rutherford was not trying to develop a new explosive when he split the atom. As J. J. Thomson said forty years ago: if the Civil Service had been in charge of research in the Stone Age, we should have had fine stone axes but would never have discovered metals.

Snobbery?

Several of those to whom I have talked blame our failure to take up new ideas on the low esteem in which engineers are allegedly held in Britain. It is true that in many Continental firms the managing director is himself a qualified engineer, able to discuss a new development in technical detail with the inventor or maker. That is rare here. And this is due, at least in part, to the poor status engineering has in the educational field. One engineer told me how, when as a boy he revealed to his headmaster his intention to become an engineer, the master ridiculed him in front of the class, saying he hoped never to hear such a thing again. John Adams, who built the great atom-smasher for C.E.R.N., takes this view also. "We produce very good physicists and very good chemists, but we don't produce very good engineers," he has said. And he adds: "The

The Denovo tyre, the "run-flat" tyre, is shown just after a charge has exploded in the tyre wall: the car drives on at high speed

first step is to improve the image of engineering at school level . . ." And the universities are not much better than the secondary schools in this respect. "When you look at engineering courses in universities in Britain, you've got to admit they are nowhere near as inspiring as pure science courses," The actual teaching is rarely done by men who are practising engineers themselves – contrary to what happens in Germany, Holland, Switzerland and elsewhere.

A report issued last year by the Science and Technology Committee of the House of Commons suggested that a special minister of state be appointed within the Department of Education and Science to concern himself with scientific and technological education. The whole subject needed much higher priority in the government's education plans, it said, and the University Grants Committee should divert funds from science to engineering if necessary. A new type of degree for applied science and engineering might be instituted.

Facing Facts

There is much going on in Britain to be proud of – as we have seen. There are many brilliant individuals, many persistent and courageous people. But we shall not escape from our current impasse merely by dwelling on our merits and achievements. It is also necessary to take a realistic view of our weaknesses and to try to correct them. It is tempting to dwell on past glories and to delude ourselves that all is still well.

For instance, we tend to think of ourselves as an exceptionally peaceful people, with an unusually low level of violence. Our police are not armed, we point out with satisfaction. But the fact is, levels of violence have been rising for decades; and though the ordinary "bobby" is not armed, the police are – have to be – issued with arms for special tasks quite routinely.

Or again, we think of Britain as a nation of craftsmen, and pride ourselves on the workmanship which goes into a Rolls-Royce, or the engraved glass of Laurence Whistler. As ship repairers we still lead the world. But the methods of mass production have tended to put craftsmanship at a discount, and the complaint too often made about British equipment is that the workmanship is shoddy and the production engineering poor.

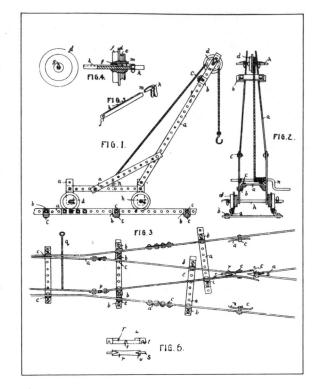

There are, to my knowledge, many workmen who would still like to turn out quality work and take a pride in it. It is up to management to find ways of giving effect to that impulse. Good workmanship is a source of deep satisfaction to the workman, quite apart from the purchaser.

There are whole areas of achievement where our performance is so poor that we ought to hang our heads in shame. Take housing. Thirty years after the war, we still have at least $1\frac{1}{2}$ million people unhoused or in substandard housing. After the war, Germany set out to build 500,000 houses and succeeded; indeed, in 1973 she built more than 700,000 houses. Britain, with a much smaller target, never achieved it in any year.

Or take the hash we have made of the nuclear power plant business, where we started ahead of the whole world.

Today our economic situation presents also a psychological danger. The emphasis is thrown heavily on making profits and on quick returns. Research is only to be backed if it promises cash return in the near future. And so on. But the greatness of a country depends very largely on enterprises which are not cost-effective – or if they are this is incidental. Hunt did not climb Mount

The 1901 patent specification for the model crane and set of points to be made from that most ingenious "toy", Meccano

Everest on a cost-effectiveness basis, and men did not reach the moon because a profit was to be made. The British achievement in the arts depends on people for whom excellence is the objective. It is worth recalling here that, in the view of many who have gone into the matter, Germany nearly won the war because of a more effective application of technology, but finally lost it because she neglected pure science.

Because many achievements of a profound nature are divorced from the profit motive, British greatness owes much to the existence of people with private incomes, people who could afford to explore, to build racing cars, to assemble art collections, to experiment. That Britain won the Battle of Britain was due, among other things, to the Spitfire, evolved by R. J. Mitchell largely with his own money, from the Supermarine S6 with which he won the Schneider Trophy three times in succession. The belief that government funding can replace private initiative is not borne out by the facts. The imaginative and determined individual will take risks which no committee would accept or even think worth accepting. Today private fortunes are few and heavily taxed, in the name of equality.

Professor G.S. Brindley plays
his modified bassoon which
produces just the same sweet
tone as the more familiar form

But the issue is a wider one. The functioning of a civilised society depends to a great extent on voluntary effort and personal altruism or sense of public obligation. Doctors meet to discuss medical matters in what is nominally their leisure time. Scientific societies produce learned journals by voluntary effort. The old are cared for by volunteers (I refer to old people who are frail but not ill) and many youth clubs are dependent on voluntary contributions and effort. The list is endless. And this is as it should be. The personal element is needed. And the recognition of a debt to society is part of civilisation; repayment of some of that debt is a source of personal satisfaction. Today, however, this network of unofficial activities is being eroded by taxation and inflation. Many learned societies can no longer afford to maintain their premises or publish their periodicals. Individuals are too hard pressed to give as much time. The state and local authorities provide some support, but it is feeble and inadequate. The deterioration of individual effort is a serious threat to British standing and achievement in the future.

It is not at all clear that greatness can be combined with egalitarianism. A society in which all are equal is by definition a society with no great men.

Our future success is also menaced by a philosophy of pleasure-seeking, or hedonism. History shows that achievement in every sphere depends on effort and self-discipline, whether it is the long hours which a ballet dancer spends at the practice-bar or the all-night sessions of an inventor. It is therefore a significant development, that some people have turned their backs on a life of high consumption to embrace the harder routines of crofting or craft-production. It is easy to dismiss such courageous individualists as eccentrics, and to confuse them with the unproductive fringe of "freaks". Perhaps we should do better to see in this trend a rebirth of the pioneering spirit, and to encourage others to do likewise.

Education does very little to help young people to arrive at a philosophy, to decide how their lives should be lived.

But if we are to emerge from the Slough of Despond, there is a requirement as important as any of these which shows little sign of being met. I mean, the emergence of inspired leadership. The British political scene has singularly failed to produce a leader who could imbue people with a sense of confidence and purpose, a leader who could set standards to which others might aspire. As Professor Arnold Toynbee pointed out, in his great *Study of History*, a civilisation is on the way out when its leaders begin to adopt the values of the masses, instead of the masses modelling themselves on the élite. Today the very idea of an élite is anathema. Even the word élite, which means those selected for their pre-eminence, is misunderstood. But that is precisely what genius is about. The future of this country depends on restoring a tradition of individual excellence; that is, on the fostering of a true élite.

But more important than all these factors I believe is the attitude of mind. Confidence in oneself and one's potentialities is half the battle. It is said that the phrase "British genius" is often greeted with a wry smile – for genius we certainly have in full measure. That is not our problem. True we find ourselves in serious financial straits and part of our problem is the economic one of finding the funds which are needed to translate the inventive genius of our countrymen into marketable products. The billion pounds that North Sea oil will earn in this and each of the next four years provide us with the needed resources, and with a breathing space during which we can plan ahead.

The one thing that remains is to recover our nerve. Dare one hope that the British Genius Exhibition will prove to be the occasion on which we ourselves recapture our belief in the British Genius?

This seagoing oceanographic buoy DB1 has been developed by the Seatek consortium (Hawker Siddeley Dynamics, E.M.I., H. & H. Green and Silley Weir); Britain will provide about ten such buoys to the E.E.C

Index